Miss W. Leggett

West Pt.

16th Jan. 167.

ELEMENTS

OF

THE CALCULUS.

ELEMENTS

OF THE

DIFFERENTIAL AND INTEGRAL

CALCULUS.

PREPARED BY

ALBERT E. CHURCH, LL.D.,

PROFESSOR OF MATHEMATICS IN THE U. S. MILITARY ACADEMY.

REVISED EDITION,

CONTAINING THE ELEMENTS OF THE

CALCULUS OF VARIATIONS.

NEW YORK:

PUBLISHED BY A. S. BARNES & BURR,

51 & 53 JOHN STREET.

1866.

RENNIE, SHEA & LINDSAY,
STEREOTYPERS AND ELECTROTYPERS,
81, 83 & 85 CENTRE-STREET,
NEW YORK.

GEORGE W. WOOD, PRINTER,
No. 2 Dutch-st., N. Y.

PREFACE.

An experience of more than twenty-five years, in teaching large classes in the U. S. Military Academy, has afforded the Author of the following pages unusual opportunities to become familiar with the difficulties encountered by most pupils, in the study of the Differential and Integral Calculus.

The results of previous endeavors to remove these difficulties, were given to the Public in former editions. Prepared, as these editions were, solely as aids to himself, in the instruction of his own pupils, he is gratified to know that they have proved acceptable to many other teachers, and that he has thus aided in extending a knowledge of this important branch of Mathematics, now absolutely necessary to thorough analytical research in the higher branches of Physical Science.

These editions have been carefully revised, and such changes introduced in the arrangement of the matter, and in the modes of demonstration, as the Author's prolonged experience has shown to be improvements. Such new matter has also been added, as he deems necessary to the perfection of the work as an elementary text-book, for those who will be satisfied with nothing short of a thorough knowledge of the subject.

The pains which have been taken to secure accuracy in the algebraic work and in the language of the text, and a clear and neat typography, will, it is hoped, render the present edition acceptable alike to teachers and pupils.

U. S. MILITARY ACADEMY,
West Point, N. Y., January 1, 1861.

CONTENTS.

PART I.

DIFFERENTIAL CALCULUS.

PART II.

INTEGRAL CALCULUS.

PART III.

CALCULUS OF VARIATIONS.

PART I.

DIFFERENTIAL CALCULUS.

Definition and Classification of Functions.

1. In the branch of Mathematics here treated, as in Analytical Geometry, two kinds of quantities are considered, viz., *variables* and *constants*; the former admitting of an infinite number of values in the same algebraic expression, while the latter admit of but one. The variables are generally designated by the last, and the constants by the first letters of the alphabet.

2. One variable quantity is a function of another, when it is so connected with it, that any change of value in the latter necessarily produces a corresponding change in the former. Thus in the expressions

$$u = bx, \qquad au^2 = cx^3,$$

u is a function of x, and x is also a function of u. Likewise, the expressions

$$ay^2, \qquad (b - y)^2,$$

are functions of y, and in each case y is a function of the expression. One of these variables is usually called the function, and the other *the independent variable*, or simply *the variable;* since to one, any arbitrary values may be *assigned*, and from the connection between the two, the corresponding values of the other *deduced.*

This relation is expressed generally thus,

$$u = f(x), \qquad u = \varphi(x), \qquad \text{or} \qquad f(u, x) = 0,$$

f and φ being mere symbols, indicating that u is a function of x. The first two expressions are read, u a function of x, or u equal to a function of x; and the third, a function of u and x equal to zero. The result, obtained by assigning a particular value to the variable, is called a state or value of the function, and each function has an infinite number of such states. Thus, if we have the function

$$(a - x)^2,$$

$$a^2, \qquad 0, \qquad a^2, \qquad 4a^2, \quad \&c.,$$

are states of the function corresponding to the particular values of x,

$$0, \qquad a, \qquad 2a, \qquad 3a, \quad \&c.$$

3. Functions are *Increasing* and *Decreasing:*

Increasing, when they are increased if the variable be increased, or decreased if the variable be decreased: *Decreasing* when they are decreased if the variable be increased, or increased if the variable be decreased. In the expressions

$$u = ax^3, \qquad u = (x + a)^3,$$

u is an increasing function of x. In the expressions

$$y = \frac{1}{x}, \qquad y = (a - x)^3,$$

y is a decreasing function of x. In the expression

$$z = (a - y)^2,$$

z is a decreasing function for all values of y less than a, but increasing for all values greater than a.

4. Functions are also *Explicit* and *Implicit:*

Explicit, when the value of the function is directly expressed in terms of the variable: *Implicit*, when this value is not directly expressed. In the examples

$$u = (a - x)^2, \qquad y = \sqrt{a^2 - x^2},$$

u and y are explicit functions of x. In the examples

$$au^2 + bx = cx^2, \qquad y^2 = a^2 - x^2,$$

or

$$au^2 + bx - cx^2 = 0, \qquad y^2 + x^2 - a^2 = 0,$$

they are implicit functions of x.

The relation between an implicit function and its variable may be expressed, either by a single equation, as above, or by two or more equations, as

$$u = ay^2, \qquad y^2 = bx,$$

in which u is an implicit function of x. The first relation is indicated generally by

$$f(u, x) = 0,$$

and the other thus.

$$u = f(y), \qquad y = \varphi(x).$$

5. Functions are also *Algebraic* and *Transcendental:*

Algebraic, when the relation between the function and variable can be expressed by the ordinary operations of Algebra, that is, by addition, subtraction, multiplication, division, the formation of powers denoted by constant exponents, and the extraction of roots indicated by constant indices: *Transcendental*, when this relation cannot be so expressed. In the examples

$$u = \log x, \qquad u = \sin(a - x), \qquad u = a^x,$$

u is a transcendental function of x. If the variable enter the exponent, the function is called *Exponential*. The logarithm of a variable expression is a *Logarithmic function*. In the expressions

$$u = \sin x, \qquad u = \cos x, \qquad u = \tan \frac{1}{x},$$

u is said to be a *Circular* function.

6. Functions are often *mixed*, being formed by the union of different kinds of simple functions, as in the expressions

$$\log x + \sin x, \qquad ax^2 + a^x,$$

7. Functions are also *Continuous* or *Discontinuous:*

Continuous, when every state obtained by substituting values of the variable between the least and greatest which give real values of the function, is real: *Discontinuous*, when any of such states are imaginary.

In the expressions

$$y = \frac{b}{a}\sqrt{a^2 - x^2}, \qquad y = \frac{b}{a}\sqrt{x^2 - a^2},$$

y in the first is continuous, in the second discontinuous; as in the one all values of x between $-a$ and $+a$ give real, while in the other they give imaginary values for y.

8. A quantity is a function of two or more independent variables, when it is so connected with them that it will change if either variable be changed, as in the examples

$$u = ax^2 + by, \qquad z = axy^2 - ux^2,$$

denoted in general thus,

$$u = f(x, y), \qquad z = F(x, y, u).$$

If in a function of a single variable, the latter be made equal to zero, the function reduces to a constant, as in the examples

$$u = ay^2, \qquad u = c + bx^2;$$

if $y = 0$, we have $u = 0$; if $x = 0$, $u = c$.

If in a function of two or more variables any one be made equal to zero, all the terms containing it will disappear, and the result will be *entirely independent of this variable*, as in the example

$$u = ax + by^2 + cz^3 + d,$$

$z = 0$ gives

$$u = ax + by^2 + d = f(x, y);$$

$z = 0$ and $y = 0$ give

$$u = ax + d = f(x).$$

If all the variables be made equal to zero, the result will be constant, as in the same example,

$z = 0$, $y = 0$, and $x = 0$, give

$$u = d = \text{a constant.}$$

Likewise, when the variable which is made equal to zero is a factor of all the terms containing any of the others, as in the example

$$u = c + ax^2y + bzy^2 = f(x, y, z),$$

$y = 0$ gives

$$u = c.$$

Definition and Properties of the Differential and Differential Coefficient.

9. To explain what is meant by *the differential of a quantity or function,* let us take the simple expression

$$u = ax^2 \ldots\ldots\ldots\ldots\ldots\ldots\ldots(1),$$

in which u is a function of x. Suppose x to be increased by another variable h; the original function then becomes $a(x+h)^2$; calling this new state of the function u', we have

$$u' = a(x+h)^2 = ax^2 + 2axh + ah^2.$$

From this, subtracting equation (1), member from member, we have

$$u' - u = 2axh + ah^2 \ldots\ldots\ldots\ldots(2).$$

The second member of this equation is the difference between the primitive and new state of the function ax^2, while h is the difference between the two corresponding states of the independent variable x. As h is entirely arbitrary, an infinite number of values may be assigned to it. Let *one* of these values, *which is to remain the same, while x is independent,* be denoted by dx, and called *differential of x,* to distinguish it from all other values of h. This particular value being substituted in equation (2), gives for the corresponding difference between the two states of u, or ax^2,

$$u' - u = 2ax \,.\, dx + a\,(dx)^2.$$

Now, *the first term of this particular difference is called the differential of u,* and is written

$$du = 2ax \,.\, dx.$$

The coefficient $(2ax)$ *of the differential of x,* in this expression, is called *the differential coefficient of the function u,* and is evidently obtained by dividing the differential of the function by the differential of the variable, and is in general written

$$\frac{du}{dx} = 2ax.$$

Resuming the expression

$$u' - u = 2axh + ah^2,$$

and dividing by h, we have

$$\frac{u' - u}{h} = 2ax + ah.$$

In the first member of this equation, the denominator is the variable increment of the variable x, and the numerator the corresponding increment of the function u; the second member is then the value of the ratio of these two increments. As h is diminished, this value diminishes and becomes nearer and nearer equal to $2ax$, and finally when $h = 0$, it becomes equal to $2ax$. From this we see, that as these increments decrease, their ratio approaches nearer and nearer to the expression $2ax$, and that by giving to h very small values, this ratio may be made to differ from $2ax$, by as small a quantity as we please. This expression is then properly, *the limit of this ratio*, and is at once obtained from the value of the ratio, by making the increment $h = 0$. It will also be seen that this limit is *precisely the same expression* as the one which we have called the differential coefficient of the function u.

What appears in this particular example is general, for let

$$u = f(x),$$

u being any function of x, and let x be increased by h, then

$$u' = f(x + h).$$

Suppose $f(x + h)$ to be developed, and arranged according to the ascending powers of h, and u to be subtracted from both members, we then have

$$u' - u = Ph + Qh^2 + Rh^3 + \&c. \dots\dots (3),$$

P, Q, R, &c., being functions of x, and every term of the second member containing h, because $u' - u$ must reduce to 0 when $h = 0$. Substituting for h the particular value dx, and taking the *first term* for the differential of u, we have

$$du = Pdx, \quad \text{and} \quad \frac{du}{dx} = P.$$

Dividing both members of equation (3) by h, we have

$$\frac{u' - u}{h} = P + Qh + Rh^2 +, \&c. \dots\dots (4).$$

Obtaining the limit of this ratio by making $h = 0$, and denoting it by L, we have

$$L = P,$$

the same value found above for $\frac{du}{dx}$; hence, *the differential coefficient of a function is always equal to the limit of the ratio of the increment of the variable to the corresponding increment of the function.*

10. The differential of a function of a single variable may then be thus defined. If the variable be increased *by a particular value, called the differential of the variable*, and the difference between the new and primitive states of the function be developed according to the ascending powers of the increment; *that term of this difference which contains the first power of the increment is the differential of the function.*

It will in general be found most convenient to obtain first the differential coefficient, for which we have the following rule:

Give to the variable a variable increment, *find the corresponding state of the function, from which subtract the primitive state, divide the remainder by the increment, obtain the limit of this ratio by making the increment equal to zero,* the result will be the differential coefficient: This, multiplied by the differential of the variable, will give the differential of the function.

The object of the Differential Calculus is, to explain the mode of obtaining and applying the differentials of functions.

11. Let the preceding principles be illustrated by the following

Examples.

1. Let $$u = bx^3.$$

For x substitute $x + h$, then,

$$u' = b\,(x + h)^3 = bx^3 + 3bx^2h + 3bxh^2 + bh^3,$$

$$u' - u = 3bx^2h + 3bxh^2 + bh^3,$$

$$\frac{u' - u}{h} = 3bx^2 + 3bxh + bh^2;$$

passing to the limit, and denoting it by L, we have

$$\text{L} = 3bx^2 = \frac{du}{dx};$$

whence

$$du = 3bx^2dx.$$

2. Let

$$u = ax^2 - cx.$$

Substituting $x + h$ for x, and subtracting, we have

$$u' - u = 2axh + ah^2 - ch,$$

$$\frac{u' - u}{h} = 2ax + ah - c;$$

making $h = 0$, we have

$$L = 2ax - c = \frac{du}{dx},$$

whence

$$du = 2axdx - cdx.$$

3. Let

$$u = \frac{a}{x},$$

then

$$u' = \frac{a}{x + h},$$

$$u' - u = \frac{a}{x + h} - \frac{a}{x} = \frac{-ah}{x^2 + xh},$$

$$\frac{u' - u}{h} = \frac{-a}{x^2 + xh},$$

and

$$L = -\frac{a}{x^2} = \frac{du}{dx};$$

whence

$$du = -\frac{adx}{x^2}.$$

4. If

$$u = 3ax^3 - mx^4, \qquad du = (9ax^2 - 4mx^3)\,dx.$$

12. Equation (4), article (9), may be put under the form

$$\frac{u' - u}{h} = \mathrm{P} + h\,(\mathrm{Q} + \mathrm{R}h + \&c.),$$

and if the expression $\mathrm{Q} + \mathrm{R}h + \&c.$ (which is a function of and h) be represented by P', this becomes

$$\frac{u' - u}{h} = \mathrm{P} + \mathrm{P}'h \ldots\ldots\ldots\ldots (1);$$

whence

$$u' = u + \mathrm{P}h + \mathrm{P}'h^2;$$

that is, the new state of the function *is equal to its primitive state, plus the differential coefficient of the function into the first power of the increment of the variable, plus a function of the variable and its increment into the second power of the increment.* This expression for the new state of the function being an important one, should be carefully remembered.

13. If we resume equation (3), Art. (9), divide by h and transpose P, we have

$$\frac{u' - u}{h} - \mathrm{P} = \mathrm{Q}h + \mathrm{R}h^2 + \&c.$$

Since when $h = 0$, the expression for the ratio $\frac{u' - u}{h}$ reduces to P, Art. (9), if h be *infinitely small*, we shall have

$$\frac{u' - u}{h} < 2\mathrm{P}, \qquad \text{or} \qquad \frac{u' - u}{h} - \mathrm{P} < \mathrm{P};$$

whence

$$\mathrm{Q}h + \mathrm{R}h^2 + \&c. < \mathrm{P},$$

and multiplying by h,

$$\mathrm{P}h > \mathrm{Q}h^2 + \mathrm{R}h^3 + \&c.$$

That is, *in a series arranged according to the ascending powers of an infinitely small quantity, the first term is numerically greater than the sum of all the others.*

14. If u be an increasing function of x, its new state u' will be greater than u, and

$$\frac{u' - u}{h} = \text{P} + \text{P}'h\ldots.\text{Art. (12)},$$

will be positive for all values of h.

If u be a decreasing function, the reverse will be the case, and the ratio be negative for all values of h.

But we see, by the preceding article, that when h is infinitely small, the sum of all the terms that follow P, in the above equation, will be less than P, and therefore the sign of P will be the same as that of the ratio; that is, *positive* when u is an increasing, and *negative* when u is a decreasing function. But P is the differential coefficient of u, Art. (9). Hence, *the differential coefficient of an increasing function is always positive; and of a decreasing function, negative.*

It should be observed, that the signs of the differential and differential coefficient are always the same.

15. Let

$$u = v,$$

u and v being functions of the variable x, which are equal to each other for every value of x. If x be increased by h, and u' and v' be the new states of u and v, we have

$$u' = v', \qquad u' - u = v' - v, \qquad \frac{u' - u}{h} = \frac{v' - v}{h}.$$

Passing to the limit of these equal ratios by making $h = 0$, we have, Art. (10),

$$\frac{du}{dx} = \frac{dv}{dx}, \qquad \text{or} \qquad du = dv;$$

that is, *if two functions of the same variable are equal, their differentials will also be equal.*

16. But if

$$u = v \pm \mathrm{C},$$

u and v being functions of x, and C a constant, and x be increased by h, we have

$$u' = v' \pm \mathrm{C}, \qquad u' - u = v' - v, \qquad \frac{u' - u}{h} = \frac{v' - v}{h},$$

and passing to the limit

$$\frac{du}{dx} = \frac{dv}{dx}, \qquad \text{or} \qquad du = d\ (v \pm \mathrm{C}) = dv;$$

that is, *if two differentials are equal, it does not follow that the expressions from which they were derived are equal.* We see also, that a constant connected by the sign $\pm$ with a variable, disappears by differentiation. In fact, *the differential of a constant is zero;* since, as it admits of no increase, there is no difference between two states, and of course no differential, Art. (10).

17. Let

$$u = \mathrm{A}v,$$

then

$$u' = \mathrm{A}v', \qquad \frac{u' - u}{h} = \mathrm{A}\frac{v' - v}{h},$$

and passing to the limit,

$$\frac{du}{dx} = A\frac{dv}{dx}, \qquad \text{or} \qquad du = d(Av) = Adv;$$

that is, *the differential of the product of a constant by a variable function, is equal to the constant multiplied by the differential of the function.*

18. When two variable quantities are so connected that one is a function of the other, either may be regarded as the function, and the other as the independent variable. Thus, from the expression $u = ax^2$, we readily obtain $x = \sqrt{\frac{u}{a}}$; in which x may be considered a function of the variable u.

In general, let

$$u = f(x) \ldots\ldots\ldots\ldots (1);$$

then by deducing the value of x,

$$x = f'(u) \ldots\ldots\ldots\ldots (2).$$

In this last expression, let the variable u be increased by any variable increment $u' - u = k$, x will receive the corresponding increment $x' - x$, and the ratio of these increments will be

$$\frac{x' - x}{k} \ldots\ldots\ldots\ldots (3).$$

If the increment $x' - x$ be denoted by h, and we substitute $x + h$ for x, in equation (1), we shall obtain, Art. (12),

$$u' - u = Ph + P'h^2 = k,$$

and substituting these values of $x' - x$ and k in expression (3), we have

$$\frac{x' - x}{k} = \frac{h}{Ph + P'h^2} = \frac{1}{P + P'h}.$$

Passing to the limit by making k, the increment of u, equal to 0, in which case $h = 0$, we have

$$\frac{dx}{du} = \frac{1}{\mathrm{P}} = \frac{1}{\frac{du}{dx}},$$

since $\mathrm{P} = \frac{du}{dx}$; that is, *the differential coefficient of x regarded as a function of u, is the reciprocal of the differential coefficient of u regarded as a function of x.*

It should be observed that du in the first member of the above equation *is constant*, u being *the independent variable*, Art. (9), while dx is variable. In the second member, the reverse is the case, dx being constant, and du variable.

To illustrate, take the example

$$u = ax^2;$$

whence

$$x = \sqrt{\frac{u}{a}}.$$

In article (9) we have found $\frac{du}{dx} = 2ax$; then

$$\frac{dx}{du} = \frac{1}{\frac{du}{dx}} = \frac{1}{2ax} = \frac{1}{2a\sqrt{\frac{u}{a}}} = \frac{1}{2\sqrt{au}}.$$

19. Let u be an implicit function of x of the second kind, Art. (4), as

$$u = f(y) \ldots\ldots\ldots (1), \qquad y = \varphi(x) \ldots\ldots\ldots (2).$$

If x be increased by h, y will receive an increment $y' - y$, which we denote by k; and these increased values of y and x in the second members of (1) and (2) will give, Art. (12),

$$u' = u + Qk + Q'k^2, \qquad y' = y + Ph + P'h^2;$$

whence

$$\frac{u' - u}{k} = Q + Q'k, \qquad \frac{y' - y}{h} = P + P'h,$$

and by multiplication,

$$\frac{u' - u}{k} \times \frac{y' - y}{h} = QP + Q'Pk + QP'h + \&c.;$$

or, since $y' - y = k$,

$$\frac{u' - u}{h} = QP + Q'Pk + QP'h + \&c.$$

Passing to the limit by making $h = 0$, which gives $k = 0$, we have

$$\frac{du}{dx} = QP.$$

But

$$Q = \frac{du}{dy}, \qquad \text{and} \qquad P = \frac{dy}{dx};$$

whence

$$\frac{du}{dx} = \frac{du}{dy} \times \frac{dy}{dx};$$

that is, *the differential coefficient of u regarded as a function of x, is equal to the differential coefficient of u regarded as a function of y, multiplied by the differential coefficient of y regarded as a function of x.*

If

$$u = f(x) \ldots\ldots\ldots (3), \qquad \text{and} \qquad v = \varphi(x) \ldots\ldots\ldots (4),$$

in which case u is evidently an implicit function of v, we find from equation (4)

$$x = \varphi'(v)\ldots\ldots\ldots\ldots(5);$$

and applying the preceding principles to equations (3) and (5), we have

$$\frac{du}{dv} = \frac{du}{dx} \times \frac{dx}{dv}\ldots\ldots\ldots\ldots(6).$$

But

$$\frac{dx}{dv} = \frac{1}{\frac{dv}{dx}}\ldots\ldots\ldots\ldots\text{Art. (18)},$$

which value in (6) gives

$$\frac{du}{dv} = \frac{\frac{du}{dx}}{\frac{dv}{dx}};$$

that is, *the differential coefficient of u regarded as a function of v, is equal to the differential coefficient of u regarded as a function of x, divided by the differential coefficient of v regarded as a function of x.*

Particular Rules for the Differentiation of Algebraic Functions.

20. In order to deduce a particular rule for the differentiation of any species of expressions, we have simply to apply to the representative of the expression, the general rule for obtaining the differential coefficient, as given in article (10), multiply by the differential of the independent variable, and then translate the result.

Let

$$u = v \pm w \pm z \ldots\ldots\ldots (1),$$

in which v, w, and z are functions of x. Increase x by h, then

$$u' = v' \pm w' \pm z';$$

subtracting (1), member from member, and dividing by h,

$$\frac{u' - u}{h} = \frac{v' - v}{h} \pm \frac{w' - w}{h} \pm \frac{z' - z}{h}.$$

Passing to the limit of these ratios, we have

$$\frac{du}{dx} = \frac{dv}{dx} \pm \frac{dw}{dx} \pm \frac{dz}{dx},$$

and multiplying by dx,

$$du = dv \pm dw \pm dz;$$

that is, *the differential of the sum or difference of any number of functions of the same variable, is equal to the sum or difference of their differentials taken separately.* Thus, if

$$u = ax^2 - bx^3,$$

$$du = d\,(ax^2) - d\,(bx^3) = 2axdx - 3bx^2dx \ldots \text{Arts. (9 \& 11)}.$$

21. Let $r = uv$ be the product of any two functions of x. If x be increased by h, we have

$$r' = u'v' = (u + Ph + P'h^2)\,(v + Qh + Q'h^2) \ldots\ldots \text{Art. (12)},$$

or performing the multiplication, subtracting the primitive product, and dividing by h,

$$\frac{r' - r}{h} = vP + uQ + \text{terms containing } h.$$

Passing to the limit,

$$\frac{dr}{dx} = vP + uQ;$$

whence

$$dr = d(uv) = vPdx + uQdx = vdu + udv,$$

since $Pdx = du$, and $Qdx = dv$, Art. (10); that is, *the differential of the product of two functions of the same variable, is equal to the sum of the products obtained by multiplying the differential of each function by the other.*

22. Let uvs be the product of three functions. Place

$$uv = r, \qquad \text{then} \qquad uvs = rs,$$

and

$$d(uvs) = d(rs) = rds + sdr \ldots\ldots\ldots (1).$$

But since

$$r = uv, \qquad dr = udv + vdu;$$

hence, by substitution in equation (1),

$$d(uvs) = uvds + sudv + svdu.$$

If we have the product of four functions $uvsw$, we may place $sw = r$, and, by a process precisely similar to the above, obtain

$$d(uvsw) = uvsdw + uvwds + uwsdv + vwsdu \ldots\ldots\ldots (2);$$

and we readily see, that by increasing the number of functions, we may in the same way prove, that *the differential of the product of any number of functions of the same variable, is equal to the sum of the products obtained by multiplying the differential of each into all the others.* Thus, if

$$uv = ax^2 . bx,$$

$$d(uv) = ax^2 . d(bx) + bx . d(ax^2) = ax^2 . bdx + bx . 2axdx = 3abx^2 dx.$$

23. If we divide both members of equation (2) of the preceding article by $uvsw$, we have

$$\frac{d(uvsw)}{uvsw} = \frac{dw}{w} + \frac{ds}{s} + \frac{dv}{v} + \frac{du}{u},$$

and we should have a similar result for any number of functions; whence we may conclude in general, that *the differential of the product of any number of functions divided by the product, is equal to the sum of the quotients obtained by dividing the differential of each function by the function itself.*

24. Let
$$u = v^m,$$

v being any function of x, and m any number, entire or fractional, positive or negative. Increase x by h, then

$$u' = v'^m = (v + Qh + Q'h^2)^m \ldots\ldots\ldots \text{Art. (12)},$$

or placing in the binomial formula,

$$(x + a)^m = x^m + max^{m-1} + \frac{m(m-1)}{1 . 2} a^2 x^{m-2} + \&c.,$$

$$v \text{ for } x, \quad \text{and} \quad (Qh + Q'h^2) \text{ for } a,$$

we have

$$u' = [v + (Qh + Q'h^2)]^m = v^m + m(Qh + Q'h^2)v^{m-1} + \&c.,$$

$$\frac{u' - u}{h} = m(Q + Q'h)v^{m-1} + \&c.,$$

each of the following terms containing h as a factor. Then

$$\frac{du}{dx} = mv^{m-1}Q,$$

$$du = dv^m = mv^{m-1}Qdx = mv^{m-1}dv \ldots\ldots\ldots (1),$$

since $Qdx = dv$, Art. (10). That is, to obtain the differential of any power of a function: *Diminish the exponent of the power by unity, and then multiply by the primitive exponent, and by the differential of the function.*

Examples.

1. If $$u = ax^4,$$

then, Art. (17),

$$du = a.dx^4 = a.4x^3dx = 4ax^3dx.$$

2. If $$u = bx^{\frac{2}{3}},$$

$$du = \frac{2}{3}bx^{\frac{2}{3}-1}dx = \frac{2}{3}bx^{-\frac{1}{3}}dx = \frac{2bdx}{3\sqrt[3]{x}}.$$

3. If $$u = cx^{-3},$$

$$du = -3cx^{-4}dx = -\frac{3cdx}{x^4}.$$

4. If
$$u = (ax - x^2)^5,$$
$$du = 5(ax - x^2)^4\, d(ax - x^2),$$

but
$$d(ax - x^2) = adx - 2xdx \ldots\ldots\ldots \text{Art. (20)};$$

hence
$$du = 5(ax - x^2)^4 (a - 2x)\, dx.$$

25. If in equation (1) of the preceding article we make $m = \frac{1}{n}$, we have
$$dv^{\frac{1}{n}} = \frac{1}{n} v^{\frac{1}{n} - 1} dv = \frac{1}{n} v^{\frac{1-n}{n}} dv = \frac{dv}{nv^{\frac{n-1}{n}}},$$

or
$$d\sqrt[n]{v} = \frac{dv}{n\sqrt[n]{v^{n-1}}}.$$

If $n = 2$, we have
$$d\sqrt{v} = \frac{dv}{2\sqrt{v}};$$

that is, the differential of a radical of the second degree, is equal to *the differential of the quantity under the radical sign divided by twice the radical.*

If $n = 3$, we have
$$d\sqrt[3]{v} = \frac{dv}{3\sqrt[3]{v^2}},$$

and in general, the differential of a radical of the nth degree, is equal to *the differential of the quantity under the radical sign divided by n times the* $(n - 1)$*th power of the radical.*

Examples.

1. If $$u = \sqrt{ax^3},$$

$$du = \frac{dax^3}{2\sqrt{ax^3}} = \frac{3ax^2dx}{2\sqrt{ax^3}} = \frac{3}{2}\sqrt{ax}.dx.$$

2. If $u = \sqrt[3]{b-x}$, $du = \dfrac{-dx}{3\sqrt[3]{(b-x)^2}}$.

3. Let $u = \sqrt[5]{bx^2}$. 4. Let $u = \sqrt{2ax - x^2}$.

26. Let $$u = \frac{s}{v} = sv^{-1},$$

s and v being functions of the same variable, then, Art. (21),

$$du = v^{-1}ds + sdv^{-1} = v^{-1}ds - sv^{-2}dv,$$

or

$$du = \frac{ds}{v} - \frac{sdv}{v^2};$$

whence, by reducing to a common denominator,

$$du = d\frac{s}{v} = \frac{vds - sdv}{v^2} \dots\dots\dots(1);$$

that is, the differential of a fraction is equal to *the denominator into the differential of the numerator, minus the numerator into the differential of the denominator, divided by the square of the denominator.*

If the denominator be constant, $dv = 0$, and equation (1) becomes

$$du = \frac{vds}{v^2} = \frac{ds}{v}.$$

If the numerator be constant, $ds = 0$, and equation (1) becomes

$$du = -\frac{sdv}{v^2}.$$

In this last case, it is evident that u is a decreasing function of v, and that its differential, when expressed in terms of dv, should be negative, Art. (14).

Examples.

1. If $$u = \frac{x}{a - x},$$

$$du = \frac{(a - x)dx - xd(a - x)}{(a - x)^2} = \frac{(a - x)dx + xdx}{(a - x)^2} = \frac{adx}{(a - x)^2}.$$

2. If $$u = \frac{ax^4}{b},$$

$$du = \frac{dax^4}{b} = \frac{4ax^3dx}{b}.$$

3. If $$u = \frac{c}{ax^3},$$

$$du = -\frac{cdax^3}{(ax^3)^2} = -\frac{3cdx}{ax^4}.$$

27. By a proper application of the preceding principles every algebraic function may be differentiated. Let them be applied to the following

Miscellaneous Examples.

1. If $$u = (a + bx^n)^p,$$

$$du = p(a + bx^n)^{p-1}d(a + bx^n)\text{........Art. (24)};$$

but
$$d(a + bx^n) = nbx^{n-1}dx;$$

hence
$$du = bnp(a + bx^n)^{p-1}x^{n-1}dx.$$

The solution of this example and many others may be simplified by applying the rule of article (19) thus; make

$$a + bx^n = z, \qquad \text{then} \qquad u = z^p,$$

$$\frac{dz}{dx} = nbx^{n-1}, \qquad \frac{du}{dz} = pz^{p-1};$$

whence

$$\frac{du}{dx} = \frac{du}{dz} \times \frac{dz}{dx} = pz^{p-1} \times nbx^{n-1} = bnp(a + bx^n)^{p-1}x^{n-1},$$

and
$$du = bnp\ (a + bx^n)^{p-1}x^{n-1}dx.$$

2. If $$u = (1 - x^2)^3,$$

$$du = 3(1 - x^2)^2d(1 - x^2) = -6\ (1 - x^2)^2xdx.$$

3. Let
$$u = \frac{ax}{x + \sqrt{a + x^2}}.$$

Place

$$y = x + \sqrt{a + x^2}, \qquad \text{then} \qquad u = \frac{ax}{y},$$

$$dy = dx + \frac{xdx}{\sqrt{a + x^2}}, \qquad du = \frac{aydx - axdy}{y^2};$$

hence

$$du = \frac{a\left\{(x + \sqrt{a + x^2})dx - x\left(dx + \frac{xdx}{\sqrt{a + x^2}}\right)\right\}}{(x + \sqrt{a + x^2})^2},$$

or, after reduction,

$$du = \frac{a^2 dx}{(x + \sqrt{a + x^2})^2 \sqrt{a + x^2}}.$$

4. If $u = \dfrac{(b + x)^2}{x}, \quad du = \dfrac{(x^2 - b^2)dx}{x^2}.$

5. $u = \sqrt[n]{a^m - x^m}, \quad du = -\dfrac{m}{n}x^{m-1}(a^m - x^m)^{\frac{1}{n}-1}dx.$

6. $u = \dfrac{2}{\sqrt[3]{a - x^2}}, \quad du = \dfrac{4}{3}x(a - x^2)^{-\frac{4}{3}}dx.$

7. $u = \dfrac{x}{x - \sqrt{1 - x^2}}, \quad du = \dfrac{-dx}{\sqrt{1 - x^2}\,(x - \sqrt{1 - x^2})^2}.$

8. Let $u = (a - \sqrt{bx^2})^3$. 9. Let $u = \dfrac{x}{(1 + x)^n}$.

10. $u = \dfrac{1 + x^2}{1 - x^2}$. 11. $u = \left(a - \sqrt{b - \dfrac{c}{x^3}}\right)^4$.

12. $u = \dfrac{\sqrt{x^2 + 1} - 1}{\sqrt{x^2 + 1} + 1}$. 13. $u = \dfrac{\sqrt{1 + x} + \sqrt{1 - x}}{\sqrt{1 + x} - \sqrt{1 - x}}$.

Successive Differentiation.

28. It is readily seen from what precedes, that the differential coefficient of a function of a single variable is, in general, a function of the same variable. It may then be differentiated, and its differential coefficient obtained.

Thus in the example,

$$u = ax^3, \qquad \frac{du}{dx} = 3ax^2 \ldots\ldots (1),$$

$3ax^2$ is a function of x, different from the primitive function.

If we differentiate both members of equation (1), we have

$$d\left(\frac{du}{dx}\right) = 6axdx.$$

But since dx is a constant, Art. (26),

$$d\left(\frac{du}{dx}\right) = \frac{d(du)}{dx} = \frac{d^2u}{dx};$$

the symbol d^2u (which is read *second differential* of u) being used to indicate that the function u has been differentiated twice, or that *the differential of the differential of* u has been taken. Hence

$$\frac{d^2u}{dx} = 6axdx, \qquad \text{or} \qquad \frac{d^2u}{dx^2} = 6ax,$$

in which dx^2 represents the square of dx, and is the same as if written $(dx)^2$.

The expression, $6ax$, being *the differential coefficient of the first differential coefficient*, is called *the second differential coefficient.*

To make the discussion general, let $u = f(x)$ and p be its differential coefficient, then

$$\frac{du}{dx} = p \ldots\ldots\ldots\ldots (2).$$

Since p is usually a function of x, let it be differentiated and its differential coefficient be denoted by q, then

$$\frac{dp}{dx} = q \ldots\ldots\ldots\ldots\ldots (3).$$

In the same way let q be differentiated and its differential co efficient be r, then

$$\frac{dq}{dx} = r \ldots\ldots\ldots\ldots\ldots (4).$$

By differentiating equation (2), we have

$$d\left(\frac{du}{dx}\right) = dp, \qquad \text{or} \qquad \frac{d^2u}{dx} = dp,$$

and by the substitution of this value of dp in (3),

$$\frac{\dfrac{d^2u}{dx}}{dx} = q, \qquad \text{or} \qquad \frac{d^2u}{dx^2} = q \ldots\ldots (5),$$

which is the second differential coefficient of the function.

By differentiating (5), we have

$$\frac{d^3u}{dx^2} = dq,$$

and by the substitution of this value of dq in (4),

$$\frac{\dfrac{d^3u}{dx^2}}{dx} = r, \qquad \text{or} \qquad \frac{d^3u}{dx^3} = r;$$

which is *the differential coefficient of the second differential coefficient*, and is called *the third differential coefficient.*

In the same way the fourth, fifth, &c., may be derived, each from the preceding, precisely as the first is obtained from the primitive function.

For this reason, the successive differential coefficients are often called, *derived functions*, and are designated thus,

$$u = f(x), \quad \frac{du}{dx} = f'(x), \quad \frac{d^2u}{dx^2} = f''(x), \text{ \&c.,}$$

$f(x)$ being the primitive function, $f'(x)$ its *first derived function*, $f''(x)$ its *second derived function*, &c.

From the differential coefficients or derived functions, we may at once obtain the corresponding differentials, by multiplying by that power of the differential of the variable, which indicates the order of the required differential, thus,

$$d^2u = \frac{d^2u}{dx^2}dx^2 = f''(x)\,dx^2,$$

$$\cdots\cdots\cdots\cdots\cdots$$

$$d^nu = \frac{d^nu}{dx^n}dx^n = f^{n\prime}(x)\,dx^n, \text{ \&c.}$$

Examples.

1. Let $$u = ax^n,$$

n being a positive whole number, then

$$\frac{du}{dx} = nax^{n-1}, \quad \frac{d^2u}{dx^2} = n(n-1)ax^{n-2},$$

$$\frac{d^3u}{dx^3} = n(n-1)(n-2)ax^{n-3},$$

$$\cdots\cdots\cdots\cdots\cdots$$

$$\frac{d^nu}{dx^n} = n(n-1)(n-2)\ldots\ldots\ldots 2.1.a.$$

Since the last differential coefficient is constant, its differential will be 0, and we have

$$\frac{d^{n+1}u}{dx^{n+1}} = 0.$$

2. Let $$u = (a - x)^{-1},$$

then

$$\frac{du}{dx} = (a - x)^{-2}, \qquad \frac{d^2u}{dx^2} = 2(a - x)^{-3},$$

. . . .

$$\frac{d^nu}{dx^n} = 2.3 \ldots\ldots n(a - x)^{-(n+1)}.$$

3. Let $$u = (a - x^2)^{\frac{3}{2}}.$$

By examining the successive differential coefficients in the above examples, it will be seen that by each differentiation the exponent of the power is diminished by unity. When this exponent is entire and positive, it will finally be reduced to 0; and, if there be no negative or fractional exponents in the expression, the corresponding differential coefficient will be constant. The next in order, as well as all which follow, will then be 0, and there will be a limited number. If the exponent be fractional, by the continued subtraction of unity the result can never be 0, but will finally, if the differentiation be continued, become negative; the successive differential coefficients will then always contain x, and there will be an infinite number. So also if the exponent be negative. And, in general, if all the exponents of an algebraic expression are entire and positive, there will be a limited number of differential coefficients. If any are negative or fractional, this number will be unlimited.

MACLAURIN'S THEOREM.

29. The object of this theorem is, *to explain the manner of developing a function of a single variable, into a series arranged according to the ascending powers of the variable with constant coefficients.*

Let $$u = f(x),$$

and let us assume a development of the proposed form,

$$u = \text{B} + \text{C}x + \text{D}x^2 + \text{E}x^3 + \&\text{c.} \ldots\ldots\ldots (1),$$

in which B, C, D, &c., are entirely independent of x, and depend upon the constants which enter into the given function. It is now required to determine such values for the constants B, C, &c., as will cause the assumed development to be a true one, for all values of x. Since these constants are independent of x, they will not change when we make $x = 0$. If then in (1) we suppose $x = 0$, and denote by A what $f(x)$ or u becomes under this supposition, we have

$$\text{A} = \text{B}.$$

Differentiating (1), and dividing by dx, we have

$$\frac{du}{dx} = \text{C} + 2\text{D}x + 3\text{E}x^2 + \&\text{c.} \ldots\ldots\ldots (2);$$

making $x = 0$, and denoting by A′ what $\frac{du}{dx}$ reduces to in this case, we have

$$\text{A}' = \text{C}.$$

Differentiating (2), and dividing by dx, we have

$$\frac{d^2u}{dx^2} = 2\text{D} + 2.3\text{E}x + \&\text{c.};$$

making $x = 0$, and denoting by A$''$ what $\frac{d^2u}{dx^2}$ becomes, we have

$$A'' = 2D; \qquad \text{whence} \qquad D = \frac{A''}{1.2}.$$

In the same way, denoting by A$'''$, A$''''$, &c., what $\frac{d^3u}{dx^3}$, $\frac{d^4u}{dx^4}$, &c., become when $x = 0$, we shall find

$$E = \frac{A'''}{1.2.3}; \qquad F = \frac{A''''}{1.2.3.4}, \text{ \&c.}$$

Substituting these values in equation (1), we have

$$u = f(x) = A + A'x + A''\frac{x^2}{1.2} \ldots\ldots\ldots + A^{n'}\frac{x^n}{1.2.3.....n} + \text{\&c.}\ldots(3),$$

in which the general term, or the one which has n terms before it, is what the nth differential coefficient of the function to be developed becomes when the variable is made equal to 0, multiplied by the nth power of the variable, and divided by the product of the consecutive numbers from 1 to n inclusive.

This formula is often written thus,

$$u = f(x) = u_{x=0} + \left(\frac{du}{dx}\right)_{x=0}\frac{x}{1}\ldots\ldots\ldots + \left(\frac{d^nu}{dx^n}\right)_{x=0}\frac{x^n}{1.2.3.....n} + \text{\&c.};$$

or

$$u = f(x) = f(0) + f'(0)x + f''(0)\frac{x^2}{1.2}\ldots\ldots\ldots + f^{n'}(0)\frac{x^n}{1.2.....n} + \text{\&c.};$$

in which the coefficients of the different powers of x are symbols denoting the same quantities as the letters A, A$'$, A$''$, &c., in formula (3).

Examples.

1. Let

$$u = (a + x)^m.$$

This, when $x = 0$, reduces to a^m; hence $A = a^m$.

By differentiation, &c., we obtain

$$\frac{du}{dx} = m(a + x)^{m-1}, \qquad \frac{d^2u}{dx^2} = m(m - 1)(a + x)^{m-2},$$

$$\frac{d^3u}{dx^3} = m(m - 1)(m - 2)(a + x)^{m-3}, \text{ \&c.}$$

Making $x = 0$ in each of these differential coefficients, we have

$$A' = ma^{m-1}, \; A'' = m(m-1)a^{m-2}, \; A''' = m(m-1)(m-2)a^{m-3}, \text{ \&c.}$$

Substituting these values in the formula (3), we have

$$(a + x)^m = a^m + ma^{m-1}x + \frac{m(m - 1)a^{m-2}x^2}{1.2} + \text{\&c.}$$

2. Let

$$u = \frac{a}{b - x} = a(b - x)^{-1}.$$

By differentiation, &c., we have

$$\frac{du}{dx} = a(b - x)^{-2} = \frac{a}{(b - x)^2}, \; \frac{d^2u}{dx^2} = 2a(b - x)^{-3} = \frac{2a}{(b - x)^3},$$

$$\frac{d^3u}{dx^3} = 2.3a(b - x)^{-4} = \frac{2.3.a}{(b - x)^4} \ldots\ldots\ldots \text{\&c.}$$

Making $x = 0$ in the original function, and in each differential coefficient, we have

$$A = \frac{a}{b}, \qquad A' = \frac{a}{b^2}, \qquad A'' = \frac{2a}{b^3} \ldots\ldots\ldots \&c.$$

These values in the formula (3) give

$$\frac{a}{b - x} = \frac{a}{b} + \frac{a}{b^2}x + \frac{a}{b^3}x^2 + \ldots\ldots\ldots \frac{a}{b^{n+1}}x^n + \ldots\ldots\ldots$$

3. Let $u = \dfrac{1}{1 + x}$. 4. Let $u = \dfrac{1}{\sqrt{1 - x^2}}$.

5. $u = \dfrac{1 + x}{1 - x}$. 6. $u = (1 + x^2)^{\frac{5}{3}}$.

Whenever the function to be developed contains the second or higher power of the variable, the work will be much abridged by substituting for this power a single variable, then making the development, and in the result resubstituting the power. Thus, in example 6, by putting z for x^2, we have

$$u = (1 + x^2)^{\frac{5}{3}} = (1 + z)^{\frac{5}{3}},$$

which is easily developed according to the ascending powers of z.

30. Functions which become infinite, when the variable on which they depend is made equal to 0; or any of the differential coefficients of which become infinite, under the same supposition, cannot be developed by Maclaurin's formula, as in such cases, either the first or some succeeding term of the series would be infinite, while the function itself would not be so.

$$u = \log x, \qquad u = \cot x, \qquad u = ax^{\frac{1}{2}},$$

are examples of such functions. In the first two A, and in the third A′, would be infinite.

Definition and Property of Functions of the Sum of two Variables.

31. A quantity is a function of the sum of two variables, when in the algebraic expression for the function, a single variable may be substituted for the sum, and the original function thus reduced, *without a change of form*, to a function of the single variable. Thus

$$u' = a(x + y)^n$$

is such a function, for if in the place of $x + y$ we substitute z, the function becomes $u' = az^n$, a function of z of the same form as the primitive function.

$$\log (x - y),$$

is also such a function of the two variables x and $-y$, which, when for $x - y$ we put z, becomes $\log z$.

If in such a function either variable be made equal to 0, the result will be a function of the other variable, of the same form as the primitive function, since the effect of this is to substitute a single variable for the sum of the other two. Thus, in the first of the above examples, if x be 0, we have

$$ay^n;$$

if y be 0, we have

$$ax^n;$$

two functions, one of y, the other of x, of the same form, which become identical if x be changed into y, or the reverse.

32. Let

$$u' = f(x + y).$$

For $x + y$ substitute z, then

$$u' = f(z).$$

If we differentiate this, first as a function of x, y being regarded as a constant; and then as a function of y, x being in turn regarded as constant, we shall have, Art. (19),

$$\frac{du'}{dx} = \frac{du'}{dz} \cdot \frac{dz}{dx}, \qquad \frac{du'}{dy} = \frac{du'}{dz} \cdot \frac{dz}{dy}.$$

But, since $z = x + y$, when y is regarded as constant, $dz = dx$; when x is constant, $dz = dy$, and the second factor in the second member of each of the above equations reduces to 1, and we have

$$\frac{du'}{dx} = \frac{du'}{dz}, \qquad \frac{du'}{dy} = \frac{du'}{dz};$$

hence

$$\frac{du'}{dx} = \frac{du'}{dy}.$$

That is, *if a function of the sum of two variables be differentiated as though one of the variables were constant, and then the same function be differentiated as though the other variable were constant, and the differential coefficients be taken, these two coefficients will be equal.*

To illustrate, let

$$u' = (x + y)^n, \quad \text{then} \quad du' = n(x + y)^{n-1} d(x + y),$$

which if y be regarded as constant becomes

$$du' = n(x + y)^{n-1} dx; \quad \text{whence} \quad \frac{du'}{dx} = n(x + y)^{n-1}.$$

And if x be regarded as constant, the same expression becomes

$$du' = n(x + y)^{n-1} dy; \quad \text{whence} \quad \frac{du'}{dy} = n(x + y)^{n-1}.$$

TAYLOR'S THEOREM.

33. The object of Taylor's Theorem is, *to explain the manner of developing a function of the algebraic sum of two variables, into a series arranged according to the ascending powers of one of the variables, with coefficients which are functions of the other and dependent also upon the constants which enter the given function.*

Let us write a development of the proposed form,

$$u' = f(x + y) = \mathrm{P} + \mathrm{Q}y + \mathrm{R}y^2 + \mathrm{S}y^3 + \&\mathrm{c.} \ldots . (1),$$

in which P, Q, R, &c., independent of y, are functions of x.

It is required to determine values for them, which substituted in equation (1) will make it true for all values of x and y. If we regard x as constant, differentiate both members of equation (1) with respect to y and divide by dy, we obtain

$$\frac{du'}{dy} = \mathrm{Q} + 2\,\mathrm{R}y + 3\,\mathrm{S}y^2 + \&\mathrm{c.}$$

If we regard y as a constant, differentiate equation (1) with respect to x and divide by dx, we obtain

$$\frac{du'}{dx} = \frac{d\mathrm{P}}{dx} + \frac{d\mathrm{Q}}{dx}y + \frac{d\mathrm{R}}{dx}y^2 + \&\mathrm{c.}$$

But by the preceding article we have $\dfrac{du'}{dy} = \dfrac{du'}{dx}$; therefore

$$\mathrm{Q} + 2\,\mathrm{R}y + 3\,\mathrm{S}y^2 + \&\mathrm{c.} = \frac{d\mathrm{P}}{dx} + \frac{d\mathrm{Q}}{dx}y + \frac{d\mathrm{R}}{dx}y^2 + \&\mathrm{c.};$$

and since, by the principle of indeterminate coefficients, the coefficients of the like powers of y in the two members must be equal,

$$\mathrm{Q} = \frac{d\mathrm{P}}{dx} \ldots\ldots\ldots (2), \quad 2\mathrm{R} = \frac{d\mathrm{Q}}{dx} \ldots\ldots\ldots 3), \quad 3\mathrm{S} = \frac{d\mathrm{R}}{dx} \ldots\ldots\ldots (4).$$

If in equation (1) we make $y = 0$; $f(x + y)$ will reduce to a function of x, Art. (31), which we denote by u. Then

$$u = \mathrm{P}.$$

Substituting this value of P in equation (2), we have

$$\mathrm{Q} = \frac{du}{dx}.$$

This value of Q in equation (3), gives

$$2\mathrm{R} = \frac{d\left(\frac{du}{dx}\right)}{dx} = \frac{d^2u}{dx^2}; \qquad \text{whence} \qquad \mathrm{R} = \frac{d^2u}{1\,.\,2\,.\,dx^2};$$

and this value of R in (4) gives

$$3\mathrm{S} = \frac{d\left(\frac{d^2u}{1\,.\,2\,.\,dx^2}\right)}{dx} = \frac{d^3u}{1\,.\,2\,.\,dx^3}; \text{ whence } \mathrm{S} = \frac{d^3u}{1\,.\,2\,.\,3\,.\,dx^3}.$$

By the substitution of these values of P, Q, R, &c., in equation (1), we have Taylor's formula;

$$u' = f(x + y) = u + \frac{du}{dx}\frac{y}{1} + \frac{d^2u}{dx^2}\frac{y^2}{1\,.\,2} + \ldots\ldots \frac{d^nu}{dx^n}\frac{y^n}{1\,.\,2\,.\,3 \ldots n} + \ldots\ldots$$

By an examination of the several terms of this formula, we see that the first (u) is what the function to be developed becomes, when the variable, according to the ascending powers of which the series is to be arranged, is made equal to 0. The second $\left(\frac{du}{dx}\frac{y}{1}\right)$ is the first differential coefficient of the first term, multiplied by the first power of this variable; and the general term is *the* nth *differential coefficient of the first term*, multiplied by the nth power of the variable, and divided by the product of the consecutive numbers from 1 to n inclusive.

The development of $f(x-y)$ is obtained from the formula by changing $+y$ into $-y$; thus

$$f(x-y) = u - \frac{du}{dx}y + \frac{d^2u}{dx^2}\frac{y^2}{1.2} - \frac{d^3u}{dx^3}\frac{y^3}{1.2.3} + \&c.$$

Examples.

1. Let

$$u' = (x+y)^m.$$

Making $y=0$, we obtain $u = x^m$, and thence by differentiation,

$$\frac{du}{dx} = mx^{m-1}, \qquad \frac{d^2u}{dx^2} = m(m-1)x^{m-2},$$

$$\frac{d^3u}{dx^3} = m(m-1)(m-2)x^{m-3}, \qquad \frac{d^nu}{dx^n} = m(m-1)\dots(m-n+1)x^{m-n}.$$

These values being substituted in the formula, give

$$u' = (x+y)^m = x^m + mx^{m-1}y + \frac{m(m-1)x^{m-2}y^2}{1.2} + \dots\dots\dots$$

$$\frac{m(m-1)\dots\dots\dots(m-n+1)x^{m-n}y^n}{1.2.3\dots\dots\dots n} + \dots\dots\dots$$

If it were required to develop the function in terms of the ascending powers of x, we should make $x=0$, and obtain y^m for the first term, from which the other terms are derived as before.

2. Let

$$u' = \frac{a}{x+y}.$$

Making $y=0$, we obtain $u = \frac{a}{x}$ for the first term; thence

$$\frac{du}{dx} = -\frac{a}{x^2}, \qquad \frac{d^2u}{dx^2} = \frac{2a}{x^3},$$

$$\frac{d^3u}{dx^3} = -\frac{2.3.a}{x^4}, \qquad \frac{d^nu}{dx^n} = \pm\frac{2.3......na}{x^{n+1}}.$$

These values being substituted in the formula, give

$$u' = \frac{a}{x+y} = \frac{a}{x} - \frac{a}{x^2}y + \frac{a}{x^3}y^2 .... \pm \frac{a}{x^{n+1}}y^n ......$$

3. Develop $u' = \frac{b}{(x-y)^{\frac{1}{3}}}$ according to the powers of $-y$.

4. Develop $u' = \frac{a}{(x-y)^2}$ according to the powers of x.

34. Since in the formula of Taylor, the coefficients of the different powers of one variable are functions of the other, it is plain that if such a value be assigned to the other, as to reduce any of these coefficients to infinity, the second member will become infinite, and the formula fail to give a development for this particular value; as, in this case, the first member will become a function of the first variable, which function is not necessarily equal to infinity for a particular value of the second variable, on which it in no way depends. Thus, in the example

$$u' = \sqrt{a+x+y},$$

which, when developed according to the ascending powers of y, gives

$$u' = \sqrt{a+x} + \frac{1}{2\sqrt{a+x}}y - \frac{1}{8\sqrt{(a+x)^3}}y^2 ....,$$

the particular value $x = -a$ reduces the coefficients of the powers of y to infinity, while the original function is reduced to $\sqrt{y}$. We should thus have $\sqrt{y} = \infty$, which cannot be. For every other value of x, however, these coefficients will be finite and the development true.

The difference between this failing case and that of Maclaurin's formula is marked. In this, the failure is only for a particular value of that variable which enters the coefficients, all other values of both variables giving a true development; while in the former case, if the formula fails to develop a function for one value of the variable, it fails for every other value.

35. If
$$u = f(x),$$

and x be increased by h, we have for the second state

$$u' = f(x + h),$$

and by changing y into h in Taylor's formula, we obtain

$$u' = f(x + h) = u + \frac{du}{dx}h + \frac{d^2u}{dx^2}\frac{h^2}{1.2} + \&c. \ldots. (1),$$

which is *the development of the second state of a function.*

Otherwise, by substituting for u, $\frac{du}{dx}$, $\frac{d^2u}{dx^2}$, &c., the expressions $f(x)$, $f'(x)$, $f''(x)$, &c., as in Art. (28), we have

$$u' = f(x + h) = f(x) + f'(x)h + f''(x)\frac{h^2}{1.2} + \&c.;$$

that is, *the new state of the function is equal to its primitive state, plus its first derived function into the first power of the increment, plus its second derived function into the second power of the increment, divided by* 1.2, *plus*, &c.; and this is but another form of Taylor's formula.

If the second state corresponding to a particular value of x, as $x = a$, be required, we have simply to substitute a for x, nd obtain

$$f(a+h) = f(a) + f'(a)h + f''(a)\frac{h^2}{1.2} + \text{\&c.};$$

in which $f(a)$, $f'(a)$, $f''(a)$, &c., are symbols denoting what the primitive function and its successive differential coefficients, or derived functions, become when a is substituted for x.

From (1) we have

$$u' - u = \frac{du}{dx}h + \frac{d^2u}{dx^2}\frac{h^2}{1.2} + \frac{d^3u}{dx^3}\frac{h^3}{1.2.3} + \text{\&c.}$$

If we now put for h the particular value dx, we have

$$u' - u = du + \frac{d^2u}{1.2} + \frac{d^3u}{1.2.3} + \text{\&c.}$$

36. If in the development of $f(x+y)$ by Taylor's formula, we suppose $x = 0$, and represent by A, A′, A″, &c., what u, $\frac{du}{dx}$, $\frac{d^2u}{dx^2}$, &c., become under this supposition, we have

$$f(y) = \text{A} + \text{A}'y + \frac{\text{A}''y^2}{1.2} + \frac{\text{A}'''y^3}{1.2.3} + \text{\&c.}$$

A, A′, A″, &c., being constant, and since y is the only variable, we may write x for it, and thus have

$$f(x) = \text{A} + \text{A}'x + \frac{\text{A}''x^2}{1.2} + \frac{\text{A}'''x^3}{1.2.3} + \text{\&c.},$$

which is identical with Maclaurin's formula.

Differentiation of Logarithmic and Exponential Functions.

37. Let $$u = \log v,$$

in which v is any function of x, and the logarithm is taken in any system. Increase x by h, then

$$u' = \log v', \qquad u' - u = \log v' - \log v = \log \frac{v'}{v}.$$

Substituting for v' its value, Art. (12), this becomes

$$u' - u = \log \frac{v + Ph + P'h^2}{v} = \log \left(1 + \frac{Ph + P'h^2}{v}\right) \ldots . (1).$$

By placing $\frac{Ph + P'h^2}{v}$ for y in the formula (Davies' Bourdon, Art. 236),

$$\log (1 + y) = M\left(y - \frac{y^2}{2} + \frac{y^3}{3} - \&c.\right),$$

we obtain

$$\log \left(1 + \frac{Ph + P'h^2}{v}\right) = M\left(\frac{Ph + P'h^2}{v} - \frac{h^2 (P + P'h)^2}{v^2} + \&c.\right);$$

and this in equation (1), after dividing by h, gives

$$\frac{u' - u}{h} = M\left(\frac{P + P'h}{v} - \frac{h (P + P'h)^2}{v^2} + \&c.\right);$$

whence, by passing to the limit,

$$\frac{du}{dx} = \mathrm{M}\frac{\mathrm{P}}{v}, \qquad du = \mathrm{M}\frac{\mathrm{P}dx}{v} = \mathrm{M}\frac{dv}{v},$$

since $\mathrm{P}dx = dv$.

For the Naperian system, $\mathrm{M} = 1$, and this expression becomes

$$d\,lu = \frac{du}{u}.\text{*}$$

The differential of the logarithm of a quantity is, then, equal to *the modulus of the system into the differential of the quantity divided by the quantity;* and this in the Naperian system, becomes *the differential of the quantity divided by the quantity.*

Examples.

1. If $$u = l\,(ax^3),$$

$$du = \frac{dax^3}{ax^3} = \frac{3\,ax^2dx}{ax^3} = 3\frac{dx}{x}.$$

2. If $$u = l\left(\frac{a}{a - x}\right),$$

$$du = \frac{d\left(\dfrac{a}{a - x}\right)}{\dfrac{a}{a - x}} = \frac{\dfrac{adx}{(a - x)^2}}{\dfrac{a}{a - x}} = \frac{dx}{a - x}.$$

The differentiation of logarithmic functions will be much simplified by the application of the principles for multiplication, division, &c., by means of logarithms. Thus, in the above example,

* Throughout the book, the symbol l before a quantity will indicate the Naperian logarithm of that quantity.

$$u = l\left(\frac{a}{a-x}\right) = la - l(a-x),$$

$$du = dla - dl(a-x) = \frac{dx}{a-x}.$$

3. Also, if $\quad u = l[(a+x)^2 \sqrt[3]{a-x}],$

$$u = 2l(a+x) + \frac{1}{3}l(a-x),$$

$$du = \frac{2dx}{a+x} - \frac{1}{3}\frac{dx}{a-x} = \frac{(5a-7x)dx}{3(a^2-x^2)}.$$

4. Let $\quad u = l\left(\frac{\sqrt{1+x^2}+x}{\sqrt{1+x^2}-x}\right).$

Multiply both terms of the fraction by the numerator; then

$$u = l(\sqrt{1+x^2}+x)^2 = 2l(\sqrt{1+x^2}+x),$$

$$du = \frac{2d(\sqrt{1+x^2}+x)}{\sqrt{1+x^2}+x} = \frac{2dx}{\sqrt{1+x^2}}.$$

5. If $\quad u = l\left(\frac{\sqrt{1+x}+\sqrt{1-x}}{\sqrt{1+x}-\sqrt{1-x}}\right), \quad du = -\frac{dx}{x\sqrt{1-x^2}}.$

6. Let $u = l\left(\frac{a-x}{x}\right)^2.$ 7. Let $u = l(a+x)^2(a-x)^2.$

8. $\quad u = l\left(\frac{\sqrt{1-x}}{\sqrt{1+x^2}}\right).$ 9. $\quad u = l(a-x^2)\sqrt{x}.$

10. Let $$u = (lx)^n;$$

then, Art. (24),

$$du = n(lx)^{n-1} dlx = \frac{n(lx)^{n-1} dx}{x}.$$

11. Let $$u = l(lx);$$

then $$du = \frac{dlx}{lx} = \frac{dx}{xlx}.$$

38. It has been seen, Art. (30), that log x cannot be developed according to the ascending powers of x. To obtain a logarithmic series, let us take $u = \log(a + x)$, and develop it by Maclaurin's formula. By differentiation, &c.,

$$du = \frac{Mdx}{a + x}; \qquad \frac{du}{dx} = \frac{M}{a + x} = M(a + x)^{-1};$$

$$\frac{d^2u}{dx^2} = -M(a + x)^{-2} = \frac{-M}{(a + x)^2}; \quad \frac{d^3u}{dx^3} = \frac{2M}{(a + x)^3}.$$

Making $x = 0$, we have for the values of A, A′, A″, &c., in the formula,

$$A = \log a, \quad A' = \frac{M}{a}, \quad A'' = -\frac{M}{a^2}, \quad A''' = \frac{2M}{a^3}, \text{ \&c.};$$

whence

$$u = \log a + M\left(\frac{x}{a} - \frac{x^2}{2a^2} + \frac{x^3}{3a^3} \cdots\cdots \pm \frac{x^n}{na^n} \cdots\cdots\right);$$

in which the logarithm of a quantity is expressed by a series, arranged according to the ascending powers of a quantity less by a.

If $a = 1$, since $\log a = \log 1 = 0$, the above series becomes,

$$u = \log(1 + x) = M\left(x - \frac{x^2}{2} + \frac{x^3}{3} \ldots\ldots \pm \frac{x^n}{n} \ldots\ldots\right),$$

the ordinary logarithmic series.

39. Let us now take the exponential function,

$$u = a^v,$$

in which v is any function of x. Taking the Naperian logarithm and differentiating both members, we have

$$lu = v\,la, \qquad \frac{du}{u} = la\,dv, \qquad du = u\,la\,dv,$$

or

$$du = da^v = a^v la\,dv;$$

that is, the differential of a constant raised to a power denoted by a variable exponent, is equal to *the power, multiplied by the Naperian logarithm of the constant into the differential of the exponent.*

Examples.

1. Let

$$u = a^{bx^2},$$

$$du = a^{bx} la\ d.bx^2, = 2ba^{bx^2} la\,xdx.$$

2. Let

$$u = a^{lx}.$$

40. If the successive differential coefficients of the function $u = a^x$ be taken, we have

$$\frac{du}{dx} = a^x la, \quad \frac{d^2u}{dx^2} = a^x(la)^2, \quad \frac{d^3u}{dx^3} = a^x(la)^3, \text{ \&c.}$$

If in the primitive function and in each of these differential coefficients we make $x = 0$, we have for the values of A, A′, &c., in Maclaurin's formula,

$$A = a^0 = 1, \quad A' = la, \quad A'' = (la)^2, \ldots\ldots \quad A^{n\prime} = (la)^n,$$

and these in the formula give

$$u = a^x = 1 + (la)x + (la)^2\frac{x^2}{1.2} + (la)^3\frac{x^3}{1.2.3} + \ldots (la)^n\frac{x^n}{1.2.3\ldots\ldots n}$$

41. By the aid of logarithms we may simplify the differentiation of complicated exponential functions. For example:

1. Let $u = z^y,$

z and y being any functions of the same variable. Take the Naperian logarithms of both members, then

$$lu = lz^y = ylz;$$

and by differentiation

$$\frac{du}{u} = dy\,lz + y\frac{dz}{z};$$

whence

$$du = u\left(dy\,lz + y\frac{dz}{z}\right) = z^y lzdy + yz^{y-1}dz,$$

which is evidently *the sum of the differentials, taken by first regarding y as the only variable, and then z.*

2. Let

$$u = a^{b^x}.$$

Taking the logarithms of both members,

$$lu = b^x la, \qquad \frac{du}{u} = la\,.\,db^x = lab^x lb\,dx,$$

$$du = a^{b^x}\,b^x la\,lb\,dx.$$

3. Let

$$u = z^{t^s},$$

then

$$lu = t^s lz, \qquad \frac{du}{u} = \frac{t^s\,dz}{z} + lz(t^s lt\,ds + st^{s-1}dt),$$

$$du = z^{t^s}\,t^s\left(\frac{dz}{z} + lz\,lt\,ds + \frac{s}{t}\,lz\,dt\right).$$

4. Let $\quad u = x^{a^x}.$ $\qquad$ 5. $\quad u = (\sqrt{\ })^x.$

6. $\qquad u = x^{\frac{1}{x}}.$ $\qquad$ 7. $\quad u = \dfrac{x}{e^x - 1}.$

Differentiation of the Circular Functions.

42. Since any arc of a circle, when less than 90°, is greater than its sine, and less than its tangent, we must have for all values of y less than 90°,

$$\frac{\sin y}{y} < 1 \qquad \text{and} \qquad \frac{\sin y}{\tan y} < \frac{\sin y}{y}.$$

But

$$\tan y = \frac{\sin y}{\cos y}; \qquad \text{whence} \qquad \frac{\sin y}{\tan y} = \cos y \ldots .(1).$$

Making $y = 0$, $\cos y$ becomes 1, and we have for the limit of the ratio (1),

$$L = 1;$$

and since $\frac{\sin y}{y}$ cannot exceed unity, nor be less than $\frac{\sin y}{\tan y}$, it must, for all small values of y, be included between them; and as $\frac{\sin y}{\tan y}$ approaches the limit 1, $\frac{\sin y}{y}$ must approach the same limit; that is, *the limit of the ratio of an arc to its sine is unity.*

43. Let

$$u = \sin x.$$

Increase x by h, then

$$u' = \sin(x + h), \qquad u' - u = \sin(x + h) - \sin x,$$

or by placing $x + h$ for p and x for q in the formula,

$$\sin p - \sin q = 2\left[\sin \tfrac{1}{2}(p - q) \cos \tfrac{1}{2}(p + q)\right],$$

$$u' - u = 2 \sin \tfrac{1}{2}h \cos(x + \tfrac{1}{2}h).$$

Dividing both members by h, and then both terms of the fraction in the second member by 2,

$$\frac{u' - u}{h} = \frac{\sin \frac{1}{2}h}{\frac{1}{2}h} \cos(x + \tfrac{1}{2}h),$$

and passing to the limit, since

$$\left(\frac{\sin \frac{1}{2}h}{\frac{1}{2}h}\right)_{h=0} = 1,*$$

$$\frac{du}{dx} = \cos x;$$

whence

$$du = d \sin x = \cos x\, dx.$$

If

$$u = \cos x,$$

$$du = d \cos x = d \sin(90° - x) = \cos(90° - x)\, d(90° - x);$$

whence

$$d \cos x = - \sin x\, dx.$$

If

$$u = \text{ver-sin}\, x,$$

$$du = d\, \text{ver-sin}\, x = d(1 - \cos x) = - d \cos x;$$

whence

$$d\, \text{ver-sin}\, x = \sin x\, dx.$$

If

$$u = \tan x,$$

$$du = d \tan x = d\, \frac{\sin x}{\cos x}$$

$$= \frac{(\cos x\, d \sin x - \sin x\, d \cos x)}{\cos^2 x} = \frac{dx(\cos^2 x + \sin^2 x)}{\cos^2 x};$$

whence

$$d \tan x = \frac{dx}{\cos^2 x}.$$

---

* This notation indicates that the expression for the quantity within the parenthesis becomes unity when $h = 0$.

If $$u = \cot x,$$

$$du = d \cot x = d \tan (90^\circ - x) = \frac{d (90^\circ - x)}{\cos^2 (90^\circ - x)};$$

whence $$d \cot x = - \frac{dx}{\sin^2 x}.$$

If $$u = \sec x,$$

$$du = d \sec x = d \frac{1}{\cos x} = \frac{\sin x \, dx}{\cos^2 x};$$

whence

$$d \sec x = \frac{\tan x \, dx}{\cos x} = \tan x \sec x \, dx.$$

If $$u = \operatorname{cosec} x,$$

$$du = d \operatorname{cosec} x = d \sec (90^\circ - x) = \cot x . \operatorname{cosec} x \, d (90^\circ - x);$$

whence $$d \operatorname{cosec} x = - \cot x . \operatorname{cosec} x \, dx.$$

If any other radius than 1 be used, it must be introduced into these formulas, by rendering them homogeneous, as in Trigonometry. Thus the formulas for the differential of the sine and cosine become

$$d \sin x = \frac{\cos x \, dx}{R}, \qquad d \cos x = - \frac{\sin x \, dx}{R}.$$

Examples.

1. If $$u = \sin \frac{bx}{a},$$

$$du = \cos \frac{bx}{a} d \frac{bx}{a} = \frac{b}{a} \cos \frac{bx}{a} dx.$$

2. If $$u = \cos\frac{1}{x},$$

$$du = -\sin\frac{1}{x}\,d\,\frac{1}{x} = \frac{1}{x^2}\sin\frac{1}{x}\,dx.$$

3. If $$u = \tan(a - x)^2,$$

$$du = \frac{d(a-x)^2}{\cos^2(a-x)^2} = -\frac{2(a-x)\,dx}{\cos^2(a-x)^2}.$$

4. If $$u = \cot^2 x,$$

$$du = 2\cot x\,d\cot x = -\frac{2\cot x\,dx}{\sin^2 x}.$$

5. If $$u = (\cos x)^{\sin x},$$

make $\cos x = z$, $\sin x = y$; then $u = z^y$, and Art. (41),

$$du = z^y lz\,dy + yz^{y-1}dz = dx\,(\cos x)^{\sin x}\left(\cos x\;l\cos x - \frac{\sin^2 x}{\cos x}\right).$$

6. Let $u = \dfrac{\sin(1+x)}{x}$. 7. Let $u = \tan(-m\sqrt{x})$.

44. In the preceding article we have found the differentials of the sine, cosine, &c., in terms of the arc as an independent variable; let it now be required to find the differential of the arc, in terms of its sine, cosine, &c.

If $u = \sin x$, then $x = \sin^{-1}u$,*

$$du = \cos x\,dx, \qquad \text{and} \qquad \frac{du}{dx} = \cos x.$$

---

* The notation $\sin^{-1}u$, $\tan^{-1}u$, &c., is used to designate *the arc whose sine is u; whose tangent is u,* &c.

If now x be regarded as the function, and u as the independent variable, we have, Art. (18),

$$\frac{dx}{du} = \frac{1}{\frac{du}{dx}} = \frac{1}{\cos x},$$

and since $\cos x = \sqrt{1 - \sin^2 x} = \sqrt{1 - u^2}$,

$$\frac{dx}{du} = \frac{1}{\sqrt{1 - u^2}}; \qquad \text{whence} \qquad dx = \frac{du}{\sqrt{1 - u^2}}.$$

If

$$u = \cos x, \qquad x = \cos^{-1} u, \qquad \frac{du}{dx} = -\sin x;$$

$$\frac{dx}{du} = -\frac{1}{\sin x} = -\frac{1}{\sqrt{1 - \cos^2 x}} = -\frac{1}{\sqrt{1 - u^2}};$$

whence

$$dx = -\frac{du}{\sqrt{1 - u^2}}.$$

If

$$u = \text{ver-sin}\, x, \qquad x = \text{ver-sin}^{-1} u,$$

$$\frac{du}{dx} = \sin x, \qquad \text{and} \qquad \frac{dx}{du} = \frac{1}{\sin x};$$

or since

$$\sin x = \sqrt{(2 - \text{ver-sin}\, x)\, \text{ver-sin}\, x} = \sqrt{(2 - u)\, u},$$

$$\frac{dx}{du} = \frac{1}{\sqrt{(2 - u)\, u}}; \qquad \text{whence} \qquad dx = \frac{du}{\sqrt{2u - u^2}}.$$

If

$$u = \tan x, \qquad x = \tan^{-1} u, \qquad \frac{du}{dx} = \frac{1}{\cos^2 x},$$

$$\frac{dx}{du} = \cos^2 x = \frac{1}{\sec^2 x} = \frac{1}{1 + \tan^2 x};$$

whence

$$dx = \frac{du}{1 + u^2}.$$

When required, the radius may be introduced into these formulas, as in the preceding article. Thus, the last formula will become

$$dx = \frac{R^2 du}{R^2 + u^2}.$$

Examples.

1. If $\quad x = \sin^{-1} 2u\sqrt{1 - u^2},$

$$dx = \frac{d(2u\sqrt{1 - u^2})}{\sqrt{1 - (2u\sqrt{1 - u^2})^2}} = \frac{2\,du}{\sqrt{1 - u^2}}.$$

2. If $\quad x = \tan^{-1}\left(-\frac{c}{y}\right),$

$$dx = \frac{d\left(-\frac{c}{y}\right)}{1 + \left(\frac{c}{y}\right)^2} = \frac{c\,dy}{c^2 + y^2}.$$

3. If $\quad u = \cos^{-1}\frac{y}{a - y}, \qquad du = \frac{-a\,dy}{(a - y)\sqrt{a^2 - 2ay}}.$

4. If $u = \text{ver-sin}^{-1}\frac{1}{x}$, $du = \frac{-dx}{x\sqrt{2x-1}}$.

45. To develop the sine and cosine of x, in terms of the ascending powers of x, we use Maclaurin's formula. Thus:

1. $$u = \sin x, \qquad \frac{du}{dx} = \cos x,$$

$$\frac{d^2u}{dx^2} = -\sin x, \qquad \frac{d^3u}{dx^3} = -\cos x, \text{ \&c.}$$

Making $x = 0$, we obtain for the values of A, A′, &c., in the formula,

$$A = 0, \quad A' = 1, \quad A'' = 0, \quad A''' = -1, \text{ \&c.};$$

thence

$$u = \sin x = \frac{x}{1} - \frac{x^3}{1.2.3} + \frac{x^5}{1.2.3.4.5} - \text{\&c.}$$

2. $$u = \cos x,$$

$$\frac{du}{dx} = -\sin x, \quad \frac{d^2u}{dx^2} = -\cos x, \quad \frac{d^3u}{dx^3} = \sin x, \text{ \&c.};$$

in which, making $x = 0$, we obtain

$$A = 1, \quad A' = 0, \quad A'' = -1, \quad A''' = 0, \text{ \&c.};$$

and thence

$$u = \cos x = 1 - \frac{x^2}{1.2} + \frac{x^4}{1.2.3.4} - \text{\&c.}$$

These series, for small values of x, are very converging, and will give with great accuracy the values of $\sin x$ and $\cos x$ for small arcs, and may therefore be used in the calculation of a table of natural sines, &c. Thus, R being unity, we have for the semi-circumference or π, the number 3,14159.....; this divided by 180, and the quotient by 60, will give the length of the arc 1′, which value, substituted for x in the series, will give the sine and cosine of one minute.

46. We can also develop the arc in terms of its sine, tangent, &c. If

$$x = \sin^{-1}u, \qquad \frac{dx}{du} = \frac{1}{\sqrt{1 - u^2}} \ldots \text{Art. (44)},$$

$$\frac{d^2x}{du^2} = u(1-u^2)^{-\frac{3}{2}}, \quad \frac{d^3x}{du^3} = (1-u^2)^{-\frac{3}{2}} + 3u^2(1-u^2)^{-\frac{5}{2}}, \text{ \&c.}$$

Making $u = 0$, we obtain

$$A = 0, \qquad A' = 1, \qquad A'' = 0, \qquad A''' = 1, \text{ \&c.};$$

and by substitution in Maclaurin's formula,

$$x = \sin^{-1}u = u + \frac{u^3}{1.2.3} + \frac{3u^5}{1.2.4.5} + \text{\&c.}$$

If $u = \frac{1}{2} = \sin 30°$, this series becomes

$$x = \sin^{-1}\frac{1}{2} = 30° = \frac{1}{2} + \frac{1}{1.2.3.2^3} + \frac{3}{1.2.4.5.2^5} + \text{\&c.},$$

by the summation of which, we find

$$30° = 0,52359\ldots\ldots,$$

and multiplying by 6, $180° = \pi = 3,14159\ldots\ldots$

47. If

$$x = \tan^{-1}u, \quad \frac{dx}{du} = \frac{1}{1 + u^2} = (1 + u^2)^{-1} \ldots\ldots \text{Art. (44)},$$

and the development may be made as in the preceding article; or otherwise thus: Developing $(1 + u^2)^{-1}$ by the binomial formula, we have

$$\frac{dx}{du} = 1 - u^2 + u^4 - u^6 + \&c. \ldots\ldots\ldots (1);$$

and since, by differentiation, the exponent of u in each term is diminished by unity, we must have, before the differentiation, an expression of the form,

$$x = Au + Bu^3 + Cu^5 + \&c.;$$

whence

$$\frac{dx}{du} = A + 3Bu^2 + 5Cu^4 + \&c. \ldots\ldots\ldots (2).$$

Comparing the coefficients of the like powers of u in (1) and (2),

$$A = 1, \; 3B = -1, \text{ and } B = -\frac{1}{3}; \; 5C = 1, \text{ and } C = \frac{1}{5}, \; \&c.;$$

whence

$$x = \tan^{-1}u = u - \frac{u^3}{3} + \frac{u^5}{5} - \frac{u^7}{7} + \&c. \ldots\ldots (3).$$

If $u = 1 = \tan 45°$, this series becomes

$$x = 45° = 1 - \frac{1}{3} + \frac{1}{5} - \frac{1}{7} + \&c.,$$

which is not sufficiently converging to enable us to determine the length of the arc with accuracy. To obviate this difficulty, we will make use of the principle that the arc 45° is equal to the arc whose tangent is $\frac{1}{2}$, plus the arc whose tangent is $\frac{1}{3}$.*

From equation (3), by the substitution of $\frac{1}{2}$ and $\frac{1}{3}$ for u, we have

$$\tan^{-1}\frac{1}{2} = \frac{1}{2} - \frac{1}{3\,.\,2^3} + \frac{1}{5\,.\,2^5} - \frac{1}{7\,.\,2^7} + \&c.,$$

$$\tan^{-1}\frac{1}{3} = \frac{1}{3} - \frac{1}{3\,.\,3^3} + \frac{1}{5\,.\,3^5} - \frac{1}{7\,.\,3^7} + \&c.;$$

hence

$$45° = \tan^{-1}\frac{1}{2} + \tan^{-1}\frac{1}{3} =$$

$$\frac{1}{2} - \frac{1}{3\,.\,2^3} + \frac{1}{5\,.\,2^5} - \&c. + \frac{1}{3} - \frac{1}{3\,.\,3^3} + \frac{1}{5\,.\,3^5} - \&c. = 0{,}78539.....,$$

which, being multiplied by 4, gives $\pi = 3{,}14159.........$

Development of the Second State of a Function of any Number of Variables.

48. Heretofore our rules for differentiation have been limited to functions of a single variable; it is now proposed to extend them to functions of any number of independent variables.

---

* To prove this principle, take the formula

$$\tan(a + b) = \frac{\tan a + \tan b}{1 - \tan a \tan b}.$$

Make $\tan a = \frac{1}{2}$, and $a + b = 45°$;

then, $\tan 45° = 1 = \dfrac{\frac{1}{2} + \tan b}{1 - \frac{1}{2}\tan b}$; whence $\tan b = \frac{1}{3}$;

hence $45° = a + b = \tan^{-1}\frac{1}{2} + \tan^{-1}\frac{1}{3}.$

Let $$u = f(x, y);$$

x and y being entirely independent of each other. *The second state* of the function will evidently be obtained by giving to both x and y variable increments. First let x receive the increment h; $f(x, y)$ then becomes $f(x + h, y)$, which (if y for a moment be regarded as constant) may be developed according to the ascending powers of h, by Taylor's formula; whence

$$f(x+h, y) = u + \frac{du}{dx}h + \frac{d^2u}{dx^2}\frac{h^2}{1.2} + \frac{d^3u}{dx^3}\frac{h^3}{1.2.3} + \&c.\ldots.(1),$$

in which $\frac{du}{dx}, \frac{d^2u}{dx^2}$, &c., are the differential coefficients of $u = f(x, y)$, taken under the supposition that x alone is variable; and are evidently all functions of x and y. If in this development we now put $y + k$ for y, we shall obtain in the first member $f(x + h, y + k)$, which is the second state of the function u. The first term of the second member (u), being a function of x and y, will, when for y we put $y + k$, become

$$f(x, y + k) = u + \frac{du}{dy}k + \frac{d^2u}{dy^2}\frac{k^2}{1.2} + \frac{d^3u}{dy^3}\frac{k^3}{1.2.3} + \&c.$$

In the same manner, $\frac{du}{dx}$, when for y we put $y + k$, may be developed, and will give, Art. (33),

$$\left(\frac{du}{dx}\right)_{y=y+k} = \frac{du}{dx} + \frac{d\left(\frac{du}{dx}\right)k}{dy} + \frac{d^2\left(\frac{du}{dx}\right)}{dy^2}\frac{k^2}{1.2} + \&c.;$$

or reducing,

$$\left(\frac{du}{dx}\right)_{y=y+k} = \frac{du}{dx} + \frac{d^2u}{dxdy}k + \frac{d^3u}{dxdy^2}\frac{k^2}{1\ 2} + \&c.$$

Also,

$$\left(\frac{d^2u}{dx^2}\right)_{y=y+k} = \frac{d^2u}{dx^2} + \frac{d^3u}{dx^2dy}k + \frac{d^4u}{dx^2dy^2}\frac{k^2}{1.2} + \&c.,$$

$$\left(\frac{d^3u}{dx^3}\right)_{y=y+k} = \frac{d^3u}{dx^3} + \frac{d^4u}{dx^3dy}k + \&c.$$

These values being substituted in the second member of (1), give for the development of the second state of a function of two variables,

$$\begin{aligned} f(x+h, y+k) = u &+ \frac{du}{dy}k + \frac{d^2u}{dy^2}\frac{k^2}{1.2} + \frac{d^3u}{dy^3}\frac{k^3}{1.2.3} + \&c., \\ &+ \frac{du}{dx}h + \frac{d^2u}{dxdy}hk + \frac{d^3u}{dxdy^2}\frac{hk^2}{1.2} + \&c., \\ &\qquad\qquad\qquad\qquad \ldots\ldots\ldots(2). \\ &+ \frac{d^2u}{dx^2}\frac{h^2}{1.2} + \frac{d^3u}{dx^2dy}\frac{h^2k}{1.2} + \&c., \\ &+ \frac{d^3u}{dx^3}\frac{h^3}{1.2.3} + \&c. \end{aligned}$$

In this development u is the original function; $\frac{du}{dy}$ is its differential coefficient taken under the supposition that y alone varies, and is called *the partial differential coefficient of the first order, taken with respect to* y; $\frac{d^2u}{dy^2}$, $\frac{d^3u}{dy^3}$, &c., are successive differential coefficients taken under the same supposition, and are called *partial differential coefficients of the second, third, &c., order, taken with respect to* y. $\frac{du}{dx}$, $\frac{d^2u}{dx^2}$, $\frac{d^3u}{dx^3}$, are obtained from the original function under the supposition that x alone varies, and are called *partial differential coefficients of the first, second, &c., order, taken with respect to* x; $\frac{d^2u}{dxdy}$ is obtained by differentiating $\frac{du}{dx}$ with respect to y and dividing the result by dy, and is called a *partial differential coefficient of the second order, taken by differentiating first with respect to x and then with respect to y*; and, in general, $\frac{d^{m+n}u}{dx^n dy^m}$ is a partial differential coefficient of the $m + n^{\text{th}}$ order, and is obtained by differentiating first n times with respect to x, and then m times with respect to y.

By an examination of these results, we see that from a function of two variables there are derived two partial differential coefficients of the first order, viz.:

$$\frac{du}{dx} \quad \text{and} \quad \frac{du}{dy};$$

three of the second order, viz.:

$$\frac{d^2u}{dx^2}, \quad \frac{d^2u}{dxdy}, \quad \frac{d^2u}{dy^2};$$

four of the third order, &c. The expressions,

$$\frac{du}{dx}dx, \quad \frac{du}{dy}dy, \quad \frac{d^2u}{dx^2}dx^2, \quad \frac{d^2u}{dxdy}dxdy, \quad \text{\&c.},$$

obtained by multiplying the several partial differential coefficients respectively by dx, dy, dx^2, $dxdy$, &c., are called *partial differentials*, and are *the results obtained by differentiating a function of two or more variables, as though, at each differentiation, all the variables but one were constant.*

49. If, instead of first increasing x by h, we increase y by k, we shall obtain

$$f(x, y+k) = u + \frac{du}{dy}k + \frac{d^2u}{dy^2}\frac{k^2}{1.2} + \frac{d^3u}{dy^3}\frac{k^3}{1.2.3} + \&c.;$$

and if in this we put $x + h$ for x, we shall evidently deduce

$$f(x+h,\ y+k) = u + \frac{du}{dx}h + \frac{d^2u}{dx^2}\frac{h^2}{1.2} + \&c.,$$

$$+ \frac{du}{dy}k + \frac{d^2u}{dydx}kh + \&c.,$$

$$+ \frac{d^2u}{dy^2}\frac{k^2}{1.2} + \&c.,$$

which development must be identical with the one in the preceding article; hence the terms containing the like powers of h and k must be equal to each other, and we must have

$$\frac{d^2u}{dxdy} = \frac{d^2u}{dydx},\quad \frac{d^3u}{dxdy^2} = \frac{d^3u}{dy^2dx},\ldots\ldots\ldots \frac{d^{n+m}u}{dx^ndy^m} = \frac{d^{n+m}u}{dy^mdx^n},$$

which shows that we shall obtain the same result, whether we differentiate first with reference to x and then with reference to y, or the reverse.

50. Let it now be required to develop the second state of the expression

$$u = x^my^n \ldots\ldots\ldots\ldots\ldots\ldots (1).$$

Differentiating with reference to x and y respectively, we obtain

$$\frac{du}{dx} = mx^{m-1}y^n \ldots\ldots (2),\qquad \frac{du}{dy} = nx^my^{n-1} \ldots\ldots (3).$$

Now differentiating (2), first with reference to x, and afterwards with reference to y, we obtain

$$\frac{d^2u}{dx^2}=m(m-1)x^{m-2}y^n\ldots.(4),\quad \frac{d^2u}{dxdy}=mnx^{m-1}y^{n-1}\ldots(5).$$

In the same manner, by differentiating (3), first with reference to x, and then with reference to y, we obtain

$$\frac{d^2u}{dydx}=mnx^{m-1}y^{n-1}=\frac{d^2u}{dxdy},\quad \frac{d^2u}{dy^2}=n(n-1)x^my^{n-2}\ldots.(6);$$

and by continuing the differentiation of (4), (5), and (6),

$$\frac{d^3u}{dx^3}=m(m-1)(m-2)x^{m-3}y^n,\ \frac{d^3u}{dx^2dy}=m(m-1)nx^{m-2}y^{n-1}, \text{\&c.}$$

Substituting these values in the formula of article (48), we have

$$(x+h)^m(y+k)^n=x^my^n+nx^my^{n-1}k+n(n-1)x^my^{n-2}\frac{k^2}{1.2}+\text{\&c.}$$
$$+mx^{m-1}y^nh+mnx^{m-1}y^{n-1}hk+\text{\&c.}$$
$$+m(m-1)x^{m-2}y^n\frac{h^2}{1.2}+\text{\&c.}$$

51. Let us now take the general case in which u is a function of any number of independent variables; that is, let

$$u=f(x,y,z,\text{\&c.}).$$

It is plain that we may deduce the development of the second state of this function in precisely the same way as in article (48), by first increasing x and y; then in the result thus obtained increasing z, and in the new result increasing one of the other variables, and so on until each shall have received an increment: We shall thus find

$$f(x+h,y+k,z+l,\text{\&c.})=f(x,y,z,\text{\&c.})+\frac{du}{dx}h+\frac{du}{dy}k+\frac{du}{dz}l+\text{\&c.}$$

Differentiation of Functions of two or more Variables.

52. If from both members of the last equation of the preceding article we subtract $f(x, y, z, \text{\&c.})$, we have

$$f(x+h, y+k, z+l, \text{\&c.}) - f(x, y, z, \text{\&c.}) = \frac{du}{dx}h + \frac{du}{dy}k + \frac{du}{dz}l, \text{\&c.},$$

plus other terms which will be of the second degree at least, with reference to the increments h, k, l, &c.; and this is the development of the difference between the new and primitive states of a function of any number of variables. If in this development we substitute for h, k, l, &c., the constants dx, dy, dz, &c., and take the sum of all the terms of the first degree with reference to these constants for the differential of the function, thus extending the definition in Art. (10), to functions of any number of variables, we have

$$du = df(x, y, z, \text{\&c.}) = \frac{du}{dx}dx + \frac{du}{dy}dy + \frac{du}{dz}dz + \text{\&c.}$$

The first member, which is the symbol for the differential of the function, is often called *the total differential* of the function, to distinguish it from the terms in the second member, each of which is a symbol for a partial differential. From this we see that the differential of a function of any number of variables is equal *to the sum of the partial differentials of the function.*

It is important, in all operations, to preserve the notation as given for the partial differentials, as we thus not only distinguish them from the total differential du, but know in each case with reference to which variable the partial differential is taken.

Examples.

1. If
$$u = ax^2y^3,$$
$$\frac{du}{dx}dx = 2axy^3dx, \qquad \frac{du}{dy}dy = 3ax^2y^2dy;$$

hence
$$du = 2axy^3dx + 3ax^2y^2dy.$$

2. If
$$u = \frac{b(a - x^2)^2}{y^2}, \quad du = -\frac{2b}{y^3}(a - x^2)\,[2xydx + (a - x^2)\,dy].$$

3. If
$$u = axy^2z^3,$$
$$du = ay^2z^3dx + 2axyz^3dy + 3axy^2z^2dz.$$

4. If $\quad u = \tan^{-1}\dfrac{x}{y}, \qquad du = \dfrac{ydx - xdy}{y^2 + x^2}.$

5. Let $\quad u = \dfrac{ay}{\sqrt{x^2 + y^2}}. \qquad$ 6. Let $\quad u = x^y.$

53. Having obtained the first differential of a function of two variables, we may from this at once derive the successive differentials. Since
$$du = \frac{du}{dx}dx + \frac{du}{dy}dy,$$
$$d^2u = d\left(\frac{du}{dx}dx\right) + d\left(\frac{du}{dy}dy\right).$$

Differentiating $\frac{du}{dx}dx$, first with reference to x, and then with reference to y, we have

$$d\left(\frac{du}{dx}dx\right) = \frac{d^2u}{dx^2}dx^2 + \frac{d^2u}{dxdy}dxdy;$$

and in the same way,

$$d\left(\frac{du}{dy}dy\right) = \frac{d^2u}{dydx}dydx + \frac{d^2u}{dy^2}dy^2;$$

whence, since

$$\frac{d^2u}{dxdy} = \frac{d^2u}{dydx} \text{.}\ldots\ldots\ldots\ldots \text{Art. (49)},$$

$$d^2u = \frac{d^2u}{dx^2}dx^2 + 2\frac{d^2u}{dxdy}dxdy + \frac{d^2u}{dy^2}dy^2.$$

Differentiating this result, since

$$d\left(\frac{d^2u}{dx^2}dx^2\right) = \frac{d^3u}{dx^3}dx^3 + \frac{d^3u}{dx^2dy}dx^2dy,$$

$$d\left(\frac{d^2u}{dxdy}dxdy\right) = \frac{d^3u}{dx^2dy}dx^2dy + \frac{d^3u}{dxdy^2}dxdy^2,$$

$$d\left(\frac{d^2u}{dy^2}dy^2\right) = \frac{d^3u}{dy^2dx}dy^2dx + \frac{d^3u}{dy^3}dy^3,$$

we derive

$$d^3u = \frac{d^3u}{dx^3}dx^3 + \frac{3d^3u}{dx^2dy}dx^2dy + \frac{3d^3u}{dxdy^2}dxdy^2 + \frac{d^3u}{dy^3}dy^3.$$

In the same way the differentials of a higher order may be derived; and in like manner we may deduce the successive differentials of a function of any number of variables.

Development of any Function of two or more Variables.

54. If in the development (2), article (48), we make both x and y equal to 0, the first member will become a function of h and k; the first term of the second member, and the different coefficients of h and k will, under the same supposition, become constants. Denoting by A what u or $f(x, y)$ becomes when x and y are made 0; by B and B′ what the partial differential coefficients of the first order; by C, C′, and C″ what those of the second order; and by D, D′, D″, and D‴ what those of the third order become under the same supposition, we obtain

$$f(h, k) = A + (Bh + B'k) + \frac{1}{1.2}(Ch^2 + 2C'hk + C''k^2)$$

$$+ \frac{1}{1.2.3}(Dh^3 + 3D'h^2k + \&c.);$$

or since we may change h and k into x and y, we have for the general development of any function of two variables,

$$f(x, y) = A + (Bx + B'y) + \frac{1}{1.2}(Cx^2 + 2C'xy + C''y^2)$$

$$+ \frac{1}{1.2.3}(Dx^3 + 3D'x^2y + \&c.).$$

If in development (2), above referred to, we make y and k each equal to 0, u becomes a function of x alone, and we have

$$f(x + h) = u + \frac{du}{dx}h + \frac{d^2u}{dx^2}\frac{h^2}{1.2} + \frac{d^3u}{dx^3}\frac{h^3}{1.2.3} + \&c.,$$

which is Taylor's formula.

In the same development, making x, y, and k, each equal to 0, and denoting by A, A′, A″, &c., what u, $\frac{du}{dx}$, $\frac{d^2u}{dx^2}$, &c., reduce to under this supposition, we obtain

$$f(h) = A + A'h + A''\frac{h^2}{1.2} + A'''\frac{h^3}{1.2.3} + \&c.;$$

or changing h into x,

$$f(x) = A + A'x + A''\frac{x^2}{1.2} + A'''\frac{x^3}{1.2.3} + \&c.,$$

which is Maclaurin's formula.

55. By making x, y, z, &c., each equal to 0, in the development of Art. (51), and then changing h, k, l, &c., into x, y, z, &c., we may deduce the development of a function of any number of variables.

Differential Equations.

56. The most general form of an equation containing the two variables x and y, is

$$f(x, y) = f'(x, y) \ldots\ldots\ldots (1).$$

Since y, in this case, is an implicit function of x, Art. (4), we may suppose its value in terms of x to be substituted in equation (1). Each member will then be an explicit function of x; and since these functions are equal, their differentials will be equal, Art. (15). Hence, to obtain *the differential equation* of a given equation containing two variables, or *the equation expressing the relation between the variables and their differentials:* Differentiate each member as a function of a single variable, and place the two results equal.

Should either member be 0 or a constant, the differential of the other will be equal to 0.

Since every term of the differential equation thus derived will contain dx or dy, we may, by transposition, place it under the form,

$$Pdx + Qdy = 0 \ldots\ldots\ldots (2);$$

from which, after dividing by dx, we may at once obtain an expression for the differential coefficient, $\frac{dy}{dx}$.

It is also manifest that the first member of the above equation (2), may be obtained by transposing all the terms of the given equation into the first member, and taking the sum of the partial differentials, as though x and y were independent variables. It should be observed, however, that owing to the relation between x and y, dy is not constant, but will in general be a function of x.

57. If an equation contain three variables, one will necessarily be a function of the other two; and each member may be regarded as a function of two independent variables, and may be differentiated as in Art. (52), and the two results placed equal to each other.

In accordance with the same principles, and in precisely the same manner, the differential equation of one containing any number of variables may be derived.

If the differential equation derived by one differentiation be again differentiated, the new differential equation will be of the *second order*; and if this be differentiated, we shall have one of the *third order*, and so on.

Examples.

1. If we take the equation of the circle

$$y^2 = R^2 - x^2 \ldots\ldots\ldots (1),$$

differentiate each member, and equate the results, we have

$$2ydy = -2xdx \ldots\ldots\ldots (2);$$

from which, after dividing by dx and $2y$, we obtain

$$\frac{dy}{dx} = -\frac{x}{y} \ldots\ldots\ldots\ldots (3).$$

Dividing equation (2) by 2, and then differentiating, x, y, and dy, being variable, we have

$$dy^2 + yd^2y = -dx^2;$$

whence

$$\frac{d^2y}{dx^2} = -\frac{1 + \frac{dy^2}{dx^2}}{y} = -\frac{1 + \frac{x^2}{y^2}}{y} = -\frac{y^2 + x^2}{y^3};$$

since $\frac{dy^2}{dx^2} = \frac{x^2}{y^2} \ldots\ldots\ldots$ equation (3).

Equivalent results may be obtained by differentiating the ex pression $y = \sqrt{R^2 - x^2}$, deduced from equation (1).

2. If $$y^2 - 2mxy + x^2 - a^2 = 0 \ldots\ldots\ldots (1),$$

$$2ydy - 2mxdy - 2mydx + 2xdx = 0 \ldots\ldots\ldots (2);$$

whence

$$\frac{dy}{dx} = \frac{my - x}{y - mx}.$$

Differentiating (2), and dividing by $2dx^2$, we obtain

$$(y - mx)\frac{d^2y}{dx^2} + \frac{dy^2}{dx^2} - 2m\frac{dy}{dx} + 1 = 0;$$

from which, after the substitution of the expression for $\frac{dy}{dx}$, we may obtain the expression for the second differential coefficient.

3. Let $$y^3 - 3axy + x^3 = 0.$$

Equations derived as above, immediately from the primitive equation by differentiation, are named *immediate differential equations.*

58. Differential equations arise, not only from simple differentiation, as in the preceding article, but from the combination of the successive immediate differential equations with each other and the primitive equation, in such a way as to eliminate certain constants, or particular functions, which enter the primitive equation. Thus, if we take the equation of the right line,

$$y = ax + b \text{..........}(1);$$

differentiate, and divide by dx, we have

$$\frac{dy}{dx} = a \text{...............}(2),$$

a result which is the same for all values of b. By the substitution of this value of a in equation (1), we have

$$ydx = xdy + bdx,$$

which is the same for all values of a.

Differentiating (2) and dividing by dx, we obtain

$$\frac{d^2y}{dx^2} = 0,$$

which is entirely independent of both a and b.

2. Take also the equation of the conic sections

$$y^2 = 2px + r^2x^2 \dots\dots (3).$$

By two differentiations, we get

$$2ydy = 2pdx + 2r^2xdx,$$

$$dy^2 + yd^2y = r^2dx^2.$$

By combining the three equations, $2p$ and r^2 may readily be eliminated, and an equation obtained which will be entirely independent of them. The result of this elimination is

$$y^2dx^2 + x^2dy^2 + yx^2d^2y - 2yxdydx = 0.$$

3. By differentiating the equation

$$y^3 - 2ax^2 + a^2 = 0,$$

and eliminating a, we obtain

$$16yx^2dx^2 - 24x^3dydx + 9y^2dy^2 = 0.$$

And, in general, all the constants of any equation may be eliminated by differentiating it as many times as there are constants. The differential equations thus obtained, with the given equation, make one more than the number of constants to be eliminated; an equation may therefore be derived which will be freed from these constants. Equations thus obtained are properly *the differential equations of the species of lines*, one of which is represented by the given equation, since, being independent of the constants, they are evidently the same for all lines of the same kind referred to the same co-ordinate axes.

4. Let

$$y = (a^2 + x^2)^{\frac{m}{n}},$$

$$dy = \frac{m}{n}(a^2 + x^2)^{\frac{m}{n}-1} 2xdx = \frac{2mx(a^2 + x^2)^{\frac{m}{n}} dx}{n(a^2 + x^2)};$$

r substituting y for $(a^2 + x^2)^{\frac{m}{n}}$, and clearing of fractions,

$$n(a^2 + x^2)dy = 2mxydx;$$

a differential equation free from the irrational function.

5. Let

$$y = a \sin x - b \cos x,$$

$$dy = a \cos xdx + b \sin xdx,$$

$$d^2y = - a \sin xdx^2 + b \cos xdx^2,$$

$$d^2y = - ydx^2;$$

which is free from the circular functions.

6. $y = e^x \cos x.$ 7. $y = l(\sin x).$

59. The Differential Calculus enables us also to eliminate, from an equation containing three variables, an arbitrary function of either two, the form of which may be entirely unknown. Thus, if

$$u = \mathrm{F}[f(x, y)],$$

the form of the function designated by the symbol F being arbitrary, we can find a differential equation expressing a relation between x, y, and the partial differential coefficients $\frac{du}{dx}$, $\frac{du}{dy}$, which will be the same, no matter what the form of the function F may be.

Make $$f(x, y) = z \ldots\ldots\ldots\ldots(1),$$

then $$u = \mathrm{F}(z).$$

Differentiating this, first with reference to x, as the independent variable, and then with reference to y, we obtain, Art. (19),

$$\frac{du}{dx} = \frac{du}{dz}\cdot\frac{dz}{dx}, \qquad \frac{du}{dy} = \frac{du}{dz}\cdot\frac{dz}{dy};$$

dividing these equations member by member, and then clearing of fractions, we have,

$$\frac{du}{dx}\cdot\frac{dz}{dy} = \frac{du}{dy}\cdot\frac{dz}{dx} \ldots\ldots\ldots(2).$$

By substituting in this the values of $\frac{dz}{dy}$ and $\frac{dz}{dx}$, taken by differentiating equation (1), we shall have the required differential equation. Such equations are called *partial differential equations.*

To illustrate, suppose

1. $f(x, y) = ax + by,$ and $u = \mathrm{F}(ax + by).$

Place $ax + by = z,$ then

$$\frac{dz}{dx} = a, \qquad \text{and} \qquad \frac{dz}{dy} = b.$$

These values in equation (2), give

$$b\frac{du}{dx} - a\frac{du}{dy} = 0.$$

2. Let

$$f(x, y) = x^2 + y^2 = z, \quad \text{and} \quad u = \mathrm{F}(x^2 + y^2).$$

Differentiating z, we find

$$\frac{dz}{dx} = 2x, \qquad \text{and} \qquad \frac{dz}{dy} = 2y.$$

These values in equation (2), give

$$y\frac{du}{dx} - x\frac{du}{dy} = 0.$$

3. Let $f(x, y) = \frac{x}{y}$, and $u = \mathrm{F}\left(\frac{x}{y}\right)$.

4. Let $u = \mathrm{F}(e^x \sin y)$.

Change of the Independent Variable.

60. In the discussion of expressions containing the successive differentials or differential coefficients of a function, it is often desirable to change the independent variable, and to regard the primitive function, or some other variable quantity, as the independent one.

This has been done in Art. (44), and is simple in cases like this, when the first differential coefficient alone is considered. Should the second differential coefficient $\frac{d^2y}{dx^2}$ enter the expression, we must remember that it was obtained, as in Art. (28), by differentiating $\frac{dy}{dx}$ as a fraction with a constant denominator, thus obtaining

$$\frac{d^2y}{dx}.$$

If we now consider both dy and dx as variable, and differentiate $\frac{dy}{dx}$ as in Art (26), we have

$$d\left(\frac{dy}{dx}\right) = \frac{dx\,d^2y - dy\,d^2x}{dx^2};$$

and this should replace $\frac{d^2y}{dx}$ whenever it enters an expression, if we desire a result in which neither dx nor dy is regarded as the independent variable; or for $\frac{d^2y}{dx^2}$ we should substitute the above expression divided by dx, that is

$$\frac{dx\,d^2y - dy\,d^2x}{dx^3} \ldots\ldots\ldots (1).$$

If we recollect, also, that $\frac{d^3y}{dx^3}$ is obtained by differentiating $\frac{d^2y}{dx^2}$ and dividing by dx, we shall obtain an expression to replace it by differentiating expression (1), without regarding any of the differentials as constant, and dividing by dx. Thus, differentiating and reducing, we have

$$\frac{(dx\,d^3y - dy\,d^3x)\,dx + 3\,(dy\,d^2x - dx\,d^2y)\,d^2x}{dx^5} \ldots . (2);$$

and in a similar way, by differentiating this expression and dividing by dx, we shall obtain an expression to replace $\frac{dy^4}{dx^4}$.

Whenever we have any expression containing $\frac{d^2y}{dx^2}$, $\frac{d^3y}{dx^3}$, &c., in which x has been regarded as the independent variable, and it is desirable to change to a more general one, in which neither x nor y is independent, we have simply to substitute for $\frac{d^2y}{dx^2}$, $\frac{d^3y}{dx^3}$, &c., the expressions (1) and (2). If in the result we desire to make y the independent variable, we must place

$$d^2y = 0, \qquad d^3y = 0, \qquad \&c.;$$

in which case the particular expressions (1) and (2) reduce to

$$-\frac{dy\,d^2x}{dx^3}\dots(1)', \qquad \frac{3\,dy\,(d^2x)^2 - dy\,dx\,d^3x}{dx^5}\dots(2)',$$

which may be used directly when we wish to change from x to y.

Example.

If we take the equation

$$y\frac{d^2y}{dx^2} + \frac{dy^2}{dx^2} + 1 = 0,$$

and substitute for $\frac{d^2y}{dx^2}$ expression (1), we have, after reduction,

$$y\,(dx\,d^2y - dy\,d^2x) + dy^2dx + dx^3 = 0,$$

in which neither x nor y is regarded as the independent variable.

If y be regarded as the independent variable, $d^2y = 0$, and we have, after dividing by dy^3 and reducing,

$$y\frac{d^2x}{dy^2} - \frac{dx^3}{dy^3} - \frac{dx}{dy} = 0.$$

61. If we have a differential equation containing x, y, and the successive differential coefficients of y; and we also have x given as a function of another variable, or x and y as functions of two other variables, and desire to deduce an expression, in the first case, independent of x and its differentials, and in the second, independent of both x and y and their differentials, we must first transform the given equation into its most general form, as indicated in the preceding article, and then deduce the values of dy, dx, d^2y, and d^2x, from the equations expressing the relation between x and y and the new variables, and substitute them in the general form.

1 Let us have

$$\frac{d^2y}{dx^2} - \frac{x}{1 - x^2}\frac{dy}{dx} + \frac{y}{1 - x^2} = 0 \ldots\ldots\ldots (1),$$

and

$$x = \cos\theta \ldots\ldots\ldots\ldots\ldots\ldots (2).$$

Substitute in (1), for $\frac{d^2y}{dx^2}$, expression (1) of the preceding article, and we obtain the general form,

$$\frac{d^2y\,dx - d^2x\,dy}{dx^3} - \frac{x}{1 - x^2}\frac{dy}{dx} + \frac{y}{1 - x^2} = 0 \ldots. (3).$$

Differentiating equation (2), regarding θ as the independent variable, we have

$$dx = -\sin\theta\,d\theta, \qquad d^2x = -\cos\theta\,d\theta^2.$$

Substituting these in (3), and recollecting that

$$1 - x^2 = 1 - \cos^2\theta = \sin^2\theta,$$

we have, after reduction,

$$\frac{d^2y}{d\theta^2} + y = 0,$$

independent of x and its differentials.

2. Let

$$z = \frac{x\,dy - y\,dx}{y\,dy + x\,dx} \ldots\ldots (1); \qquad \left.\begin{array}{l} x = r\cos v \\ y = r\sin v \end{array}\right\} \ldots\ldots (2).$$

Differentiating equations (2), we have

$$dx = \cos v\,dr - r\sin v\,dv, \quad dy = \sin v\,dr + r\cos v\,dv.$$

Substituting these in (1), and reducing, we obtain

$$z = \frac{rdv}{dr},$$

in which either r or v may be regarded as the independent variable.

Vanishing Fractions.

62. In the discussion of the results obtained by the application of the Calculus, we often meet with expressions which, for a particular value of the variable, become $\frac{0}{0}$. This, although in general the algebraic symbol of an indeterminate quantity, does not indicate such a quantity in the particular cases referred to. As in the example

$$\frac{ax - x^2}{a^2 - x^2},$$

which becomes $\frac{0}{0}$ when $x = a$; if we divide both numerator and denominator by the common factor $a - x$, we obtain

$$\frac{x}{a + x},$$

and this, when $x = a$, reduces to $\frac{1}{2}$, which is the true value of the fraction in the particular case.

Expressions of this kind are called *vanishing fractions*, and reduce to $\frac{0}{0}$ in consequence of the existence of a factor common to both terms; which factor becomes 0 under the particular supposition.

All such fractions may be represented generally by the expression

$$\frac{\mathrm{P}(x - a)^m}{\mathrm{Q}(x - a)^n};$$

in which P and Q are functions of x, not containing the factor $(x - a)$.

There are three cases:

1. When $m = n$, the fraction becomes

$$\frac{P(x-a)^m}{Q(x-a)^m} = \frac{P}{Q};$$

which, when $x = a$, becomes a finite quantity,

$$\frac{P_{x=a}}{Q_{x=a}}.$$

2. When $m > n$, it may be put under the form

$$\frac{P(x-a)^{m-n}}{Q},$$

$m - n$ being positive; and this, when $x = a$, becomes

$$\frac{0}{Q_{x=a}} = 0.$$

3. When $m < n$, the fraction may be put under the form

$$\frac{P}{Q(x-a)^{n-m}},$$

$n - m$ being positive; and this, when $x = a$, becomes

$$\frac{P_{x=a}}{0} = \infty.$$

We see from the above, that if we can, by any process, ascertain the relative values of m and n, we shall know the true value of the fraction when $x = a$.

63. Whenever the common factor can be readily discovered, the simplest method of obtaining the true value of the fraction is to strike it out, and then put for the variable its particular value. But as in most cases it is not easy to detect this factor, other methods become necessary.

Let $\frac{r}{s}$ be a vanishing fraction, r and s being functions of x, and let a be the particular value which, substituted for x, reduces the fraction to $\frac{0}{0}$.

It is plain that, if we substitute $a + h$ for x, and, after reduction, make $h = 0$, it will amount only to the substitution of a for x. Suppose this substitution made, and that in the result both numerator and denominator are arranged so that the exponents of h shall increase from left to right; we then have

$$\left(\frac{r}{s}\right)_{x=a+h} = \frac{Ah^m + Bh^{m'} + \text{\&c.}}{A'h^n + B'h^{n'} + \text{\&c.}},$$

in which A, A′, B, B′, m, n, &c., are constants. After reducing this fraction to its lowest terms, by dividing both numerator and denominator by that power of h which is indicated by the smallest exponent, we shall have one of three cases.

1. If $m = n$,

$$\left(\frac{r}{s}\right)_{x=a+h} = \frac{A + Bh^{m'-m} + \text{\&c.}}{A' + B'h^{n'-n} + \text{\&c.}}.$$

2. If $m > n$,

$$\left(\frac{r}{s}\right)_{x=a+h} = \frac{Ah^{m-n} + \text{\&c.}}{A' + \text{\&c.}}.$$

3. If $m < n$,

$$\left(\frac{r}{s}\right)_{x=a+h} = \frac{A + \text{\&c.}}{A'h^{n-m} + \text{\&c.}}.$$

Now making $h = 0$, we have for the true value in the three cases,

$$1. \quad \left(\frac{r}{s}\right)_{x=a} = \frac{A}{A'}. \qquad 2. \quad \left(\frac{r}{s}\right)_{x=a} = \frac{0}{A'} = 0.$$

$$3. \quad \left(\frac{r}{s}\right)_{x=a} = \frac{A}{0} = \infty.$$

Whence we derive the general rule: *For the variable, substitute that value which causes the fraction to reduce to $\frac{0}{0}$, plus an increment; reduce the result to its simplest form, and then make the increment equal to* 0. The final result will be the true value of the fraction for the particular value of the variable, and may be finite, zero, or infinite.

The effect of the application of this rule is evidently, by the reduction of the fraction to its lowest terms, to cause the common factor to disappear. To illustrate, take the fraction

$$\frac{(x^2 - a^2)^{\frac{3}{2}}}{(x - a)^{\frac{3}{2}}},$$

which becomes $\frac{0}{0}$ when $x = a$.

For x, put $a + h$; the primitive fraction then becomes

$$\frac{(2ah + h^2)^{\frac{3}{2}}}{h^{\frac{3}{2}}}.$$

Dividing both terms by $h^{\frac{3}{2}}$, we obtain

$$(2a + h)^{\frac{3}{2}};$$

which, when $h = 0$, becomes $(2a)^{\frac{3}{2}}$, the true value.

In this case the common factor $(x - a)^{\frac{3}{2}}$ is evident; striking it out, we have

$$(x + a)^{\frac{3}{2}},$$

which becomes $(2a)^{\frac{3}{2}}$, when $x = a$.

64. The application of the general rule of the preceding article which is strictly algebraic, will, in most cases, give rise to complicated algebraic work. We may, however, by the aid of the Differential Calculus, deduce a practical rule of much more easy application. Thus, if the vanishing fraction, as in the preceding article, be

$$u = \frac{r}{s}, \qquad \text{then} \qquad r = us,$$

$$dr = uds + sdu;$$

in which, if we make $x = a$, we shall have (since $s_{x=a} = 0$),

$$(dr)_{x=a} = (uds)_{x=a};$$

whence

$$u_{x=a} = \left(\frac{r}{s}\right)_{x=a} = \frac{(dr)_{x=a}}{(ds)_{x=a}} \ldots\ldots\ldots (1),$$

for the true value of the fraction in the particular case.

If $(dr)_{x=a} = 0$, this value is 0.

If $(ds)_{x=a} = 0$, it is ∞.

If both are 0 at the same time, the second member of (1) becomes $\frac{0}{0}$, and $\frac{dr}{ds}$ is a new vanishing fraction; then, as above, we take the differentials of both its terms, put a for x, and thus obtain

$$u_{x=a} = \frac{(d^2r)_{x=a}}{(d^2s)_{x=a}}.$$

If this again becomes $\frac{0}{0}$, we continue the same process, and have

$$u_{x=a} = \frac{(d^3r)_{x=a}}{(d^3s)_{x=a}},$$

and so on. The rule may then be thus enunciated: *Take the differentials of the numerator and denominator; in each, substitute that value of the variable which reduces the original fraction to* $\frac{0}{0}$; *if both do not reduce to* 0 *or infinity, what the former becomes divided by what the latter becomes, will be the true value of the fraction. If both reduce to* 0, *take the second differentials, and make the same substitution; or continue the differentiation, &c., until two differentials of the same order are obtained, both of which do not become* 0 *or infinity; what one becomes divided by what the other becomes, will be the true value of the fraction.*

It should be observed, that the effect of the application of this rule is, at each differentiation, to diminish by unity the exponent of the factor which causes the fraction to reduce to $\frac{0}{0}$, Art. (28). If the exponents of this factor in the numerator and denominator are fractional, and not contained between the same two consecutive whole numbers, it is plain that the least one will be reduced to a negative number, by a less number of differentiations than will be required by the other. The differential of that term of the fraction which contains it, will then, by the substitution of the particular value of the variable, reduce to infinity, while that of the other reduces to 0, and the true value of the fraction will be either

$$\frac{\infty}{0} = \infty, \qquad \text{or} \qquad \frac{0}{\infty} = 0.$$

If, however, these exponents are contained between the same two consecutive whole numbers, they will become negative by the same number of differentiations, and the differentials of both terms of the fraction reduce to infinity at the same time, as will the successive differentials. In this, *the only failing case* of the rule, we shall not be able, by its application, to obtain the true value of the fraction, but must fall back upon the general rule, Art. (63). As an illustration of this, we may refer to the example in article (63), in which the second differentials, and all which follow, become infinite when $x = a$.

Examples.

1. If

$$\frac{r}{s} = \frac{x^n - 1}{x - 1},$$

which becomes $\frac{0}{0}$ when $x = 1$,

$$dr = nx^{n-1}dx, \qquad ds = dx;$$

making $x = 1$, in each of these, we have

$$(dr)_{x=1} = ndx, \qquad (ds)_{x=1} = dx,$$

and

$$\left(\frac{r}{s}\right)_{x=1} = \frac{ndx}{dx} = n.$$

2. If

$$\frac{r}{s} = \frac{1 - \sin x}{\cos x},$$

which becomes $\frac{0}{0}$ when $x = \frac{\pi}{2}$,

$$dr = -\cos x dx, \qquad ds = -\sin x dx;$$

making $x = \frac{\pi}{2}$ in each, we have

$$(dr)_{x=\frac{\pi}{2}} = 0, \qquad (ds)_{x=\frac{\pi}{2}} = -dx,$$

the quotient of which is 0, the true value of the fraction.

3. If

$$\frac{r}{s} = \frac{ax^2 - 2acx + ac^2}{bx^2 - 2bcx + bc^2},$$

$$dr = (2ax - 2ac)\,dx, \qquad ds = (2bx - 2bc)\,dx,$$

both of which reduce to 0, when $x = c$. Differentiating again,

$$d^2r = 2adx^2, \qquad d^2s = 2bdx^2,$$

and

$$\left(\frac{r}{s}\right)_{x=c} = \frac{a}{b}.$$

4. Take $\dfrac{a^x - b^x}{x}$ when $x = 0$. Ans. $la - lb$.

5. $\dfrac{ml\left(1 + \frac{x}{a}\right)}{x}$ $x = 0$. Ans. $\dfrac{m}{a}$.

6. $\dfrac{1 - \sin x + \cos x}{\sin x + \cos x - 1}$ $x = \dfrac{\pi}{2}$.

7. $\dfrac{a - x - ala + alx}{a - \sqrt{2ax - x^2}}$ $x = a$.

8. $\dfrac{x^x - x}{1 - x + lx}$ $x = 1$.

9. $\dfrac{x - 2\sin x}{x\sin x}$ $x = 0.$

10. $\dfrac{x - \sin x}{x^3}$ $x = 0.$

11. $\dfrac{x^2}{1 - \cos mx}$ $x = 0.$

65. We sometimes meet with the product of two factors, one of which becomes 0, and the other ∞, for a particular value of the variable. Let rt be such a product, in which r becomes 0, and t infinite. It may be written

$$rt = \frac{r}{\frac{1}{t}},$$

which, for the particular value, becomes $\frac{0}{0}$. Its value may then be determined as in the preceding articles.

Example.

Let $rt = (1 - x)\tan\dfrac{\pi x}{2},$ when $x = 1.$

Writing it under the proposed form, we have

$$rt = \frac{1 - x}{\dfrac{1}{\tan\dfrac{\pi x}{2}}} = \frac{1 - x}{\cot\dfrac{\pi x}{2}},$$

the true value of which, when $x = 1$, is $\dfrac{2}{\pi}$.

66. The fraction $\frac{r}{s}$ may become $\frac{\infty}{\infty}$, in which case it may be written

$$\frac{r}{s} = r \times \frac{1}{s},$$

which becomes $\infty \times \frac{1}{\infty} = \infty \times 0$, and may then be treated as in the preceding article.

67. Sometimes, also, we find expressions which become $\infty - \infty$.

Let $$\frac{1}{r} - \frac{1}{s}$$

be such an expression, r and s becoming 0. It may be written

$$\frac{1}{r} - \frac{1}{s} = \frac{s - r}{rs},$$

which will reduce to $\frac{0}{0}$. For an example, take

$$\frac{x}{\cot x} - \frac{\pi}{2 \cos x},$$

which becomes $\infty - \infty$, when $x = \frac{\pi}{2}$. By reduction we obtain

$$\frac{x \sin x - \frac{\pi}{2}}{\cos x};$$

the true value of which is -1, when $x = \frac{\pi}{2}$.

Maxima and Minima of Functions of a single Variable.

68. Let $u = f(x)$, and suppose x to be increased by insensible degrees from its least value, until we obtain a corresponding state of the function which is greater than the state which immediately precedes it, and greater also than that which immediately follows it; this state of the function is called *a maximum*. If we obtain a state which is less than both of these consecutive states, it is *a minimum*. We say, then, that a function of a single variable is at a maximum state, or *a maximum, when it is greater than the state which immediately precedes, and greater also than the state which immediately follows it;* and *a minimum, when it is less than both of these states.*

69. If u is a function of x, and x supposed to be increasing, it is evident that when passing from the preceding states to its maximum, u must *increase* as x increases, that is, be an *increasing function* of x; and when passing from its maximum to the succeeding states, it must *decrease* as x increases, that is, be *a decreasing function* of x. In the first case, Art. (14), the sign of its first differential coefficient must be positive, and in the second, negative; therefore at the maximum state *the first differential coefficient must change its sign from plus to minus, as the variable increases.* For a similar reason at a minimum state, the first differential coefficient must change its sign *from minus to plus;* and these changes of sign, in the first differential coefficient, are respectively the *analytical characteristics* of the maximum and minimum states of a function. But, as a function which is continuous can change its sign only by becoming zero or infinity, it follows that no value of the variable will give a maximum or minimum value to the function, unless the same value reduces the first differential coefficient to zero or infinity.

The roots of the two equations,

$$\frac{du}{dx} = 0 \ldots . (1), \qquad \text{and} \qquad \frac{du}{dx} = \infty, \text{ or } \frac{dx}{du} = 0 \ldots . (2),$$

will then give all the values of x which can possibly make u a maximum or a minimum. After having obtained these roots, let each, first with an infinitely small decrement and then with an infinitely small increment, be substituted in the given function; the results will be the states which immediately precede and follow the one obtained by substituting the root itself; if both are less than this, the latter will be a maximum; if both are greater, a minimum.

Or, as will in general be more convenient, let each of these roots, with an infinitely small decrement and increment, be successively substituted in the first differential coefficient; if the first result be positive, and the second negative, the root will make the function a maximum; if the reverse, a minimum. If the two results have the same sign, the root under consideration will give neither a maximum nor a minimum.

Since equations (1) and (2) may give several roots which will fulfil the required conditions, there may be more than one maximum or minimum state of the same function; and, therefore, the maximum state is not necessarily the greatest state, nor the minimum the least.

Examples.

1. If $$u = a + (x - b)^2 \ldots\ldots\ldots\ldots\ldots (3),$$

$$\frac{du}{dx} = 2(x - b), \qquad \text{and} \qquad \frac{dx}{du} = \frac{1}{2(x - b)}.$$

Placing $\frac{du}{dx} = 0$, we have

$$2(x - b) = 0; \qquad \text{whence} \qquad x = b.$$

If in equation (3) we substitute first $b - h$ for x, and then $b + h$, denoting the corresponding states of the function by u'' and u', we have

$$u'' = a + h^2, \qquad \text{and} \qquad u' = a + h^2,$$

both of which are *greater* than $u = a$, the result obtained by substituting b for x; *hence* $u = a$ *is a minimum.*

The only value of x which will reduce $\frac{dx}{du}$ to 0, is $x = \infty$; there is then no finite value of x which will satisfy this condition, hence $x = b$ gives the only minimum state, and there is no maximum.

2. If $$u = a - (x - b)^{\frac{2}{3}} \ldots\ldots.(4),$$

$$\frac{du}{dx} = \frac{-2}{3(x - b)^{\frac{1}{3}}}, \qquad \text{and} \qquad \frac{dx}{du} = \frac{-3(x - b)^{\frac{1}{3}}}{2}.$$

Placing $\frac{du}{dx} = 0$, we obtain $x = \infty$, which gives no finite solution.

Placing $\frac{dx}{du} = 0$, we have

$$3(x - b) = 0; \qquad \text{whence} \qquad x = b.$$

If, then, in (4), we substitute first $b - h$, and then $b + h$, for x, we have

$$u'' = a - h^{\frac{2}{3}}, \qquad \text{and} \qquad u' = a - h^{\frac{2}{3}},$$

both of which are less than $u = a$, the result of the substitution of b for x; $u = a$ *is then a maximum and the only one, and there is no minimum.*

If, in the first differential coefficients in the above examples, we substitute $b - h$ and $b + h$ for x, we obtain in the first, for $b - h$ a negative, and for $b + h$ a positive result; and in the second the reverse, as it should be.

70. When the states which immediately precede and follow the maximum or minimum state of u, can be deduced from Taylor's formula, a more convenient practical rule may be applied. To demonstrate it, let

$$u = f(x);$$

then let $x + h$ be substituted for x, and the difference between the two states be developed, as in Art. (35), and we shall have

$$u' - u = \frac{du}{dx}h + \frac{d^2u}{dx^2}\frac{h^2}{1.2} + \frac{d^3u}{dx^3}\frac{h^3}{1.2.3} + \&c.\ldots.(1).$$

If, in this, h be infinitely small and negative, u' will be the state immediately preceding u; and if h be positive, u' will be the state immediately following u; and in both cases, the first term of the second member will be greater numerically than the sum of all the others, Art. (13), and the sign of the second member will be the same as that of its first term. Now, if u be a maximum, it must be greater than u', whether h be positive or negative; that is, $u' - u$ must be negative in both cases; and if u be a minimum, $u' - u$ must be positive. But $\frac{du}{dx}h$ evidently changes its sign as h changes from negative to positive; u cannot, therefore, be either a maximum or a minimum, unless the term $\frac{du}{dx}h$ disappears, which, since h is not zero, requires that

$$\frac{du}{dx} = 0 \ldots\ldots\ldots\ldots\ldots\ldots.(2).$$

The roots of this equation will then, in the case under consideration, give all the values of x which can possibly make u either a maximum or minimum.

Let a be one of these roots; to ascertain whether it will make u a maximum or minimum, substitute it in equation (1), and we have, since $\left(\frac{du}{dx}\right)_{x=a} = 0,$

$$(u' - u)_{x=a} = \left(\frac{d^2u}{dx^2}\right)_{x=a} \frac{h^2}{1.2} + \left(\frac{d^3u}{dx^3}\right)_{x=a} \frac{h^3}{1.2.3} + \&c.\ldots(3).$$

The sign of the second member will now be the same as $\left(\frac{d^2u}{dx^2}\right)_{x=a}$, since h^2 is positive. If $\left(\frac{d^2u}{dx^2}\right)_{x=a}$ is negative, then u will be greater than u', whether h be positive or negative, and $u_{x=a}$ will be a maximum. If $\left(\frac{d^2u}{dx^2}\right)_{x=a}$ be positive, $u_{x=a}$ will be a minimum. If $\left(\frac{d^2u}{dx^2}\right)_{x=a} = 0$, then the sign of (3) will depend upon the sign of $\left(\frac{d^3u}{dx^3}\right)_{x=a} \frac{h^3}{1.2.3}$, which evidently changes its sign as h changes; and there can be neither a maximum nor minimum for $x = a$, unless $\left(\frac{d^3u}{dx^3}\right)_{x=a} = 0.$

In this case the sign will depend upon that of $\left(\frac{d^4u}{dx^4}\right)_{x=a}$, and there will be a maximum when this is negative, and a minimum when it is positive, and so on; if the first differential coefficient which does not reduce to 0, is of an odd order, there will be no maximum nor minimum for $x = a$; if of an even order, there will be one or the other, according as its sign is negative or positive. If the first differential coefficient which does not reduce to 0, becomes infinity, this is a failing case of Taylor's formula, Art. (34), and the rule thus demonstrated fails with it. Whence, to determine the maximum or minimum states of a given function: *Find its first differential coefficient and place it equal to 0; substitute each of the real roots of the equation thus formed, in the second differential coefficient. Each one which gives a negative result will, when substituted in the function, make it a maximum; and each which gives a positive result, will make it a minimum. If either reduce the second differential coefficient to 0, substitute in the third, fourth, &c., until one be obtained which does not reduce to 0. If this be of an odd order, the root will correspond to neither a maximum nor minimum; if of an even order and negative, there will be a corresponding maximum; if positive, a minimum. Substitute the root in the function; the result will be the corresponding maximum or minimum.*

To illustrate, take the example

$$u = \frac{x^3}{3} + ax^2 - 3a^2x,$$

$$\frac{du}{dx} = x^2 + 2ax - 3a^2, \qquad \frac{d^2u}{dx^2} = 2x + 2a \ldots . (1).$$

Placing the expression for $\frac{du}{dx} = 0$, we have

$$x^2 + 2ax - 3a^2 = 0,$$

the roots of which are $x = a$, and $x = -3a$. The first substituted in (1) gives $4a$, which being positive, indicates a minimum. The second substituted in (1) gives $-4a$, which indicates a maximum. Substituting the roots in the given function, we have for the minimum $u = -\frac{5a^3}{3}$, and for the maximum $u = 9a^3$.

Abbreviations in the Application of the Rules for Maxima and Minima.

71. Let
$$v = Au,$$

u being any function of x, and A a positive constant. By differentiation, &c., we have

$$\frac{dv}{dx} = A\frac{du}{dx}, \qquad \frac{d^2v}{dx^2} = A\frac{d^2u}{dx^2};$$

from which it appears that those values of x, which make $\frac{du}{dx} = 0$, will also make $\frac{dv}{dx} = 0$, and the reverse. Also, that any of these values which will make $\frac{d^2u}{dx^2}$ negative, will make $\frac{d^2v}{dx^2}$ negative; and any which will make $\frac{d^2u}{dx^2}$ positive, will make $\frac{d^2v}{dx^2}$ positive. Hence every value of x which will make u a maximum or minimum, will make v or Au a maximum or minimum. *Therefore a constant positive factor may be omitted during the search for those values of the variable corresponding to a maximum or minimum.*

To illustrate, take the example

$$v = \frac{2bx^4 + a^3bx}{a^2} = \frac{b}{a^2}(2x^4 + a^3x).$$

Omitting the constant factor, we may write

$$u = 2x^4 + a^3x,$$

$$\frac{du}{dx} = 8x^3 + a^3, \qquad \frac{d^2u}{dx^2} = 24x^2 \ldots\ldots\ldots (1).$$

Placing the expression for $\frac{du}{dx} = 0$, we have

$$8x^3 + a^3 = 0; \qquad \text{whence} \qquad x = -\frac{a}{2}.$$

This value in (1) gives $6a^2$, and indicates a minimum, which is

$$u = -\frac{3a^4}{8}; \qquad \text{whence} \qquad v = -\frac{3a^2b}{8}.$$

72. Let $$v = u^n,$$

u and v being functions of x, and n entire. Then

$$\frac{dv}{dx} = nu^{n-1}\frac{du}{dx},$$

$$\frac{d^2v}{dx^2} = nu^{n-1}\frac{d^2u}{dx^2} + n(n-1)u^{n-2}\frac{du^2}{dx^2}.$$

Now every value of x which will make $\frac{du}{dx} = 0$, will also make $\frac{dv}{dx} = 0$; and if the same value makes nu^{n-1} *positive*, it will give to $\frac{d^2v}{dx^2}$ the same sign as $\frac{d^2u}{dx^2}$ (since $\frac{du^2}{dx^2} = 0$); that is, if it makes u a maximum or minimum, it will make v a maximum or minimum. If it makes nu^{n-1} *negative*, it will give to $\frac{d^2v}{dx^2}$ a sign contrary to that of $\frac{d^2u}{dx^2}$; that is, if it makes u a maximum, it will make v a minimum, and the reverse.

All values of x, however, which will make $v = u^n$ a maximum or minimum, will not necessarily make u a maximum or minimum, for the equation

$$\frac{dv}{dx} = nu^{n-1}\frac{du}{dx} = 0,$$

may be satisfied by making either

$$nu^{n-1} = 0, \qquad \text{or} \qquad \frac{du}{dx} = 0.$$

Those values of x which satisfy the first, and not the second of these equations, will make u neither a maximum nor minimum, but may make $v = u^n$ a maximum or minimum. As in the example

$$v = (a^3 - x^3)^2 = u^2,$$

$$dv = 2udu, \qquad \frac{dv}{dx} = 2u\frac{du}{dx}.$$

We may make $\frac{dv}{dx} = 0$, by placing either

$$2u = 2(a^3 - x^3) = 0, \qquad \text{whence} \qquad x = a;$$

or $$\frac{du}{dx} = -3x^2 = 0, \qquad \text{whence} \qquad x = 0.$$

The value $x = a$ evidently makes v a minimum, but as it does not reduce $\frac{du}{dx} = -3x^2$ to 0, it will make u neither a maximum nor minimum.

The value $x = 0$ answers to neither a maximum nor a minimum. As the corresponding power of a radical expression is formed by *omitting the radical sign, we may*, in accordance with the above principles, *omit it, and seek those values of the variable which will make the power a maximum or minimum.* We are sure thus to get all the values which will make the root a maximum or minimum. Care should be taken, however, not to use any of those which belong only to the power.

To illustrate, take the example

$$u = \sqrt[3]{ax^3 - 3x^2}.$$

Omitting the radical sign, we have

$$v = ax^3 - 3x^2.$$

Taking the first differential coefficient of v and placing it equal to 0, we find the two roots, $x = 0$ and $x = \frac{2}{a}$. The first gives a maximum value, 0, for both v and u. The second gives a minimum value for both v and u, viz.:

$$v = -\frac{4}{a^2}, \qquad u = \sqrt[3]{-\frac{4}{a^2}}.$$

It would not be proper to extract the root of a function before applying the rule, as those values which make the power, and not the root, a maximum or minimum, would thus be excluded.

73. In a manner similar to the above, it may be shown that any value of the variable which will render u a maximum or minimum, will also render $\log u$ and a^u a maximum or minimum; and also any value which will make u a maximum, will make $\frac{1}{u}$ a minimum, and the converse.

74. It often happens that the first differential coefficient is composed of two or more variable factors, each of which, when placed equal to 0, will give real roots of the equation $\frac{du}{dx} = 0$. In this case we may easily ascertain what the second differential coefficient reduces to, by the substitution of any one of these roots, without deducing the expression for the second differential coefficient itself. Thus, let

$$\frac{du}{dx} = XX'$$

be such a coefficient, X being 0 when $x = a$. Then

$$\frac{d^2u}{dx^2} = X\frac{dX'}{dx} + X'\frac{dX}{dx};$$

or, since $X = 0$ when $x = a$,

$$\left(\frac{d^2u}{dx^2}\right)_{x=a} = \left(X'\frac{dX}{dx}\right)_{x=a}.$$

That is, to obtain the corresponding value of the second differential coefficient: *Multiply the differential coefficient of that factor which is* 0, *by the other factors, and then substitute the particular value of the variable.* To illustrate, let

$$u = x^2(x-a)^2,$$

$$\frac{du}{dx} = 2x(x-a)(2x-a),$$

which is equal to 0, when

$$2x = 0; \qquad \text{whence} \qquad x = 0 \ldots\ldots (1).$$

$$(x-a) = 0; \qquad \text{"} \qquad x = a \ldots\ldots (2).$$

$$(2x-a) = 0; \qquad \text{"} \qquad x = \frac{a}{2} \ldots\ldots (3).$$

Taking the first factor, $2x$, and multiplying its differential coefficient by the other factors, we obtain the expression

$$2(x-a)(2x-a);$$

from which, by making $x = 0$, we obtain

$$\left(\frac{d^2u}{dx^2}\right)_{x=0} = 2a^2,$$

which indicates a minimum.

Multiplying the differential coefficient of the second factor by the other factors, and making $x = a$, we obtain $2a^2$, which indicates a minimum.

Proceeding in the same way with the third factor, we obtain $-a^2$, which indicates a maximum.

75. As the principles and rules for maxima and minima, in the preceding articles, are demonstrated independently of the nature or kind of the function, they are equally applicable to all kinds of functions of a single variable. To apply them to an *implicit function*, we have then only to find its first and successive differential coefficients, by the rules previously given, Arts. (19) and (56), and then proceed precisely as in the foregoing examples.

To illustrate, take the example

$$y^2 - 2mxy + x^2 - a^2 = 0 \ldots\ldots (1),$$

and let it be required to find the value of x which will make y a maximum or minimum. By differentiating as in article (56), we obtain

$$2ydy - 2mxdy - 2mydx + 2xdx = 0;$$

whence

$$\frac{dy}{dx} = \frac{my - x}{y - mx} \ldots\ldots\ldots\ldots (2).$$

Placing this equal to 0, we have

$$my - x = 0; \qquad \text{whence} \qquad x = my,$$

which, in equation (1), gives

$$y = \frac{a}{\sqrt{1 - m^2}}; \qquad \text{whence} \qquad x = \frac{ma}{\sqrt{1 - m^2}}.$$

Differentiating the factor $my - x$, equation (2), dividing by dx, and multiplying by $\frac{1}{y - mx}$, Art. (74), we obtain the expression

$$\frac{1}{y - mx}\left(m\frac{dy}{dx} - 1\right),$$

which, by the substitution of the values of y and x (since then $\frac{dy}{dx} = 0$), becomes

$$\frac{-1}{a\sqrt{1 - m^2}},$$

and indicates a maximum.

Solution of Practical Problems in Maxima and Minima of Functions of one Variable.

76. The only difficulty in the application of the preceding principles to the solution of problems, consists in obtaining a convenient algebraic expression for the function whose maximum or minimum state is required. No general rule can well be given by which this expression can be found. In order to indicate as clearly as possible the methods to be pursued, we will give, in detail, the solution of several cases differing from each other.

1. Required the dimensions of the maximum cylinder which can be inscribed in a given right cone.

Suppose a cylinder inscribed, as represented in the figure. Let

$$\text{VA} = a, \quad \text{BA} = b, \quad \text{VC} = x, \quad \text{CO} = y;$$

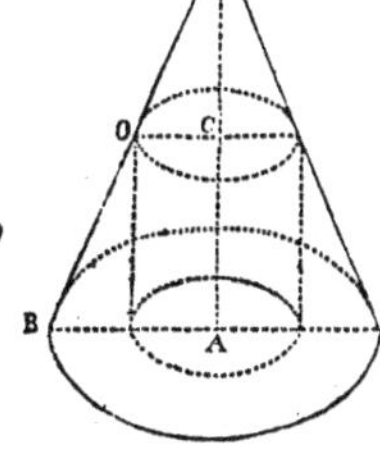

then $\text{AC} = a - x$, and the volume of the cylinder, which we denote by v, is equal to

$$\pi y^2 (a - x) \ldots\ldots (1).$$

From the similar triangles VCO and VAB, we have the proportion

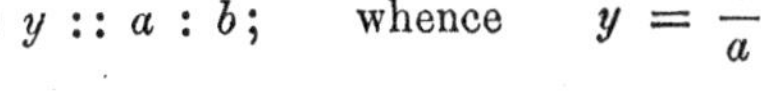

$$x : y :: a : b; \quad \text{whence} \quad y = \frac{bx}{a}.$$

Substituting this value in (1), we have

$$v = \frac{\pi b^2}{a^2} x^2 (a - x) \ldots\ldots\ldots\ldots (2).$$

Omitting the constant factor, Art. (71), we may write

$$u = ax^2 - x^3;$$

whence

$$\frac{du}{dx} = 2ax - 3x^2, \qquad \frac{d^2u}{dx^2} = 2a - 6x \ldots\ldots (3).$$

Placing $\frac{du}{dx} = 0$, we find the roots, $x = 0$ and $x = \frac{2}{3}a$. The second value of x in (3) gives $-2a$, and therefore will make v a maximum, which is $\frac{4\pi ab^2}{27}$.

For the altitude of the maximum cylinder, we have $a - x = \frac{1}{3}a$, and for the radius of the base, $y = \frac{2}{3}b$.

The first value of x in (3) gives $2a$, which indicates a minimum, which is evidently $v = 0$.

2. Required to draw a tangent to the given quadrant ABD, so that the triangle CFG shall be a minimum.

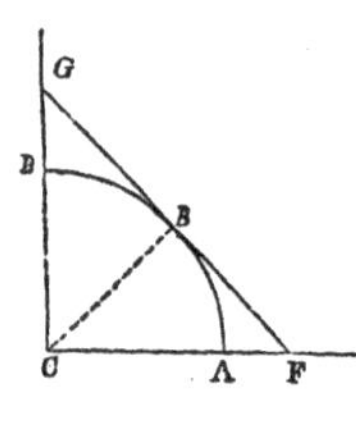

Let $CB = R$, $FB = x$, $BG = y$; then $FG = x + y$. The area of the triangle is equal to $\frac{1}{2}CB \times FG$, which, since $\frac{1}{2}CB$ is constant, will be a minimum when FG is a minimum, Art. (71). In the right-angled triangle CFG, since CB is perpendicular to FG, we have

$$R^2 = xy; \qquad \text{whence} \qquad y = \frac{R^2}{x},$$

and

$$FG = u = x + \frac{R^2}{x},$$

$$\frac{du}{dx} = 1 - \frac{R^2}{x^2} = \frac{x^2 - R^2}{x^2},$$

which, being placed equal to 0, gives $x = R$, and $y = R$.

Hence the angle $BCF = 45°$. Obtaining the corresponding value of $\frac{d^2u}{dx^2}$, as in Art. (74), we find for a result $\frac{2}{R}$.

3. The whole surface of a right cylinder being given, it is required to find the radius of the base and the altitude, when the volume is a maximum.

Let m^2 = the surface, x = the radius of the base, and z = the altitude; then

$$v = \pi x^2 z.$$

But

$$m^2 = 2\pi xz + 2\pi x^2; \qquad \text{whence} \qquad z = \frac{m^2 - 2\pi x^2}{2\pi x};$$

therefore

$$v = \frac{m^2 x}{2} - \pi x^3,$$

and $x = \sqrt{\frac{m^2}{6\pi}}$, and $z = 2\sqrt{\frac{m^2}{6\pi}}$, when v is a maximum.

4. Required the minimum distance from a point without a given circle to the circumference.

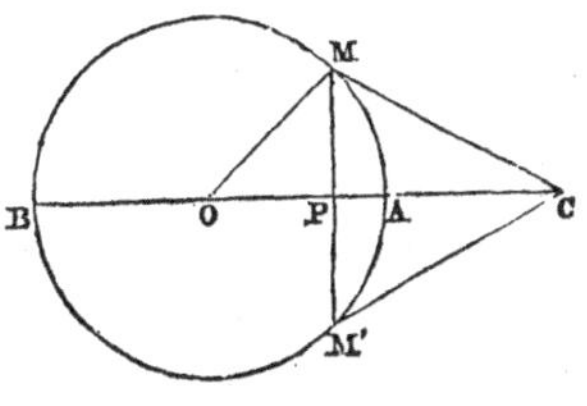

Let O be the centre of the given circle, OM its radius = R, C the given point, CO = a. Join C and O, and take OC as the axis of X, the origin being at O. Denote the co-ordinates of M by x and y, and the distance CM by u; then

$$u = \sqrt{R^2 + a^2 - 2ax}.$$

Omitting the radical sign, we have

$$u^2 = v = R^2 + a^2 - 2ax \dots\dots (1).$$

$$\frac{dv}{dx} = -2a,$$

which cannot be 0 or ∞. We should not, therefore, conclude that CM admits of no minimum value; for it is evident, from the inspection of the figure, that CA, corresponding to the value $x = R$, is a minimum. Cases of this kind are remarkable, but readily explained by a reference to either demonstration of the rule; for it depends entirely upon the principle, that the function is expressed in terms of a variable which admits of a value both *less* and *greater* than the one which corresponds to a maximum or minimum. Now, in the case under consideration, there is no value of x greater than R, which corresponds to a real state of the function. Both rules must therefore fail in cases of this kind. The remedy is, to deduce an expression for the function, in terms of some other variable which will admit of proper values. Thus, if the value of $x = \sqrt{R^2 - y^2}$ be substituted in equation (1) we have

$$v = R^2 + a^2 - 2a\sqrt{R^2 - y^2},$$

$$\frac{dv}{dy} = \frac{2ay}{\sqrt{R^2 - y^2}}, \qquad 2ay = 0, \qquad y = 0;$$

which gives $x = R$, and $x = -R$. The first value corresponds to a minimum, and the second to a maximum value of CM.

The same result might have been obtained directly, by taking any other line than CO as the axis of X, in which case x would have admitted of proper values.

5. Required the minimum isosceles triangle circumscribing a given circle.

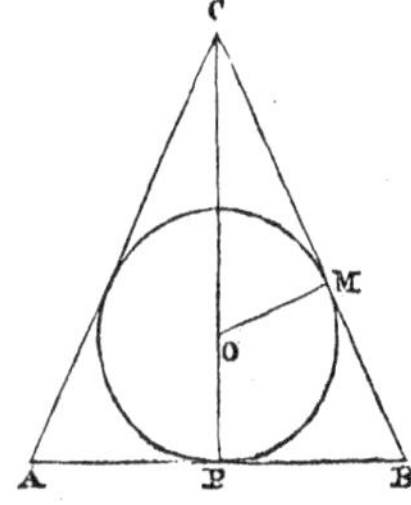

Let O be the centre of the circle, ABC the circumscribing triangle, AC and BC being the equal sides, OM $= R$, PC $= x$, AB $= 2y$, and denote the area by u; then

$$u = xy \ldots\ldots\ldots (1).$$

From the similar triangles COM and CPB, we have

$$PB : OM :: CB : CO,$$

or

$$y : R :: \sqrt{x^2 + y^2} : x - R;$$

whence

$$(x - R)y = R\sqrt{x^2 + y^2},$$

and

$$y = R\sqrt{\frac{x}{x - 2R}}.$$

Substituting this in (1), we have

$$u = \mathrm{R}x\sqrt{\frac{x}{x - 2\mathrm{R}}} = \mathrm{R}\sqrt{\frac{x^3}{x - 2\mathrm{R}}}.$$

Omitting the constant factor R, and the radical sign, Arts. (71) nd (72), we have

$$v = \frac{x^3}{x - 2\mathrm{R}}, \qquad \frac{dv}{dx} = \frac{(x - 2\mathrm{R})\,3x^2 - x^3}{(x - 2\mathrm{R})^2}.$$

Placing $\frac{dv}{dx} = 0$,

$$(x - 2\mathrm{R})\,3x^2 - x^3 = 0, \qquad 2x^3 - 6\mathrm{R}x^2 = 0;$$

whence

$$x = 0, \qquad x = 3\mathrm{R}.$$

The value $x = 3\mathrm{R}$ gives a minimum, which is the circumscribed equilateral triangle. The value $x = 0$ corresponds to no maximum nor minimum, as there is *no real state* of the function *immediately* following the state corresponding to $x = 0$; all states, from $x = 0$ to $x = 2\mathrm{R}$, being imaginary.

6. Required to divide a given quantity, a, into two parts, such that the mth power of one, multiplied by the nth power of the other, shall be a maximum.

If $x =$ one of the parts, then $\qquad x = \frac{ma}{m + n}.$

7. In a given triangle, it is required to inscribe a maximum rectangle.

The altitude of the rectangle $= \frac{1}{2}$ altitude of triangle.

8. A certain quantity of water being given, it is required to find the relation between the radius of the base and the altitude of a cylindrical vessel, open at the top, which shall just hold the water and have its interior surface a minimum.

The radius = the altitude.

9. Required the maximum rectangle which can be inscribed in a circle.

Each side $= R\sqrt{2}$.

10. Required the maximum cone which can be inscribed in a given sphere.

11. Required the minimum cone circumscribing a given sphere.

12. Required the minimum triangle formed by the axis, the produced ordinate of the extreme point, and a tangent to the arc of a parabola.

13. Required the maximum cylinder that can be inscribed in a given ellipsoid of revolution.

14. Required the axis of the maximum parabola that can be cut from a given right cone.

15. Required the minimum value of y in the equation $y = x^x$.

Maxima and Minima of Functions of two or more Variables.

77. A function of two or more variables is a maximum when it is greater, and a minimum when it is less, than all of its consecutive states. Let

$$u = f(x, y), \qquad \text{then} \qquad u' = f(x + h, y + k),$$

$$u' - u = h(p + p't) + \frac{h^2}{1.2}(q + 2q't + q''t^2) + \&c.\dots(1);$$

after placing in the development of article (48),

$$k = ht, \qquad \frac{du}{dx} = p, \qquad \frac{du}{dy} = p',$$

$$\frac{d^2u}{dx^2} = q, \qquad \frac{d^2u}{dxdy} = q', \qquad \frac{d^2u}{dy^2} = q'', \&c.$$

The sign of this series, when h is infinitely small, will depend upon the sign of its first term. We shall obtain all of the consecutive states of u by giving to h and k proper infinitely small values, both positive and negative; and therefore, when u is either a maximum or a minimum, the sign of $u' - u$ for all these values of h and k must be the same. But the first term of the series (1) evidently changes its sign when the sign of h changes; there can, then, be neither a maximum nor a minimum, unless

$$h(p + p't) = 0, \qquad \text{or} \qquad p + p't = 0;$$

and since this must be 0 for all values of $t = \frac{k}{h}$, we must have separately

$$p = 0, \qquad \text{and} \qquad p' = 0,$$

or $$\frac{du}{dx} = 0\dots(2), \qquad \frac{du}{dy} = 0\dots(3).$$

The values of x and y, deduced from these equations and substituted in the second term of series (1), (h and k being infinitely small), should make it negative for a maximum, and positive for a minimum. This term may be put under the form

$$\frac{h^2q''}{1.2}\left(\frac{q}{q''} + \frac{2q'}{q''}t + t^2\right),$$

which, if there be a maximum or minimum, must not change its sign for any value of t; that is, the quantity within the parenthesis must be positive for all values of t. By adding and subtracting $\frac{q'^2}{q''^2}$, we may write

$$\frac{q}{q''} + \frac{2q'}{q''}t + t^2 = \left(t + \frac{q'}{q''}\right)^2 + \frac{q}{q''} - \frac{q'^2}{q''^2} \ldots (4)$$

If $\frac{q}{q''} - \frac{q'^2}{q''^2}$, which is entirely independent of t, be negative, such values may be assigned to t as to make expression (4) either positive or negative. To render the entire expression positive, q and q'' must then have the same sign, and

$$\frac{q}{q''} - \frac{q'^2}{q''^2} > 0, \quad \text{or} \quad = 0,$$

that is, we must have

$$qq'' - q'^2 > 0, \qquad \text{or} \qquad qq'' - q'^2 = 0.$$

The conditions then are

$$\left(\frac{d^2u}{dxdy}\right)^2 < \frac{d^2u}{dx^2} \times \frac{d^2u}{dy^2}, \quad \text{or} \quad = \frac{d^2u}{dx^2} \times \frac{d^2u}{dy^2};$$

and also that $\frac{d^2u}{dx^2}$ and $\frac{d^2u}{dy^2}$ have the same sign, *after the values of x and y deduced from the equations* $\frac{du}{dx} = 0$ *and* $\frac{du}{dy} = 0$ *have been substituted.* And since the sign of the second term will then depend upon q'', the sign of $\frac{d^2u}{dy^2}$ must be negative for a maximum, and positive for a minimum.

If the second term becomes 0, we must substitute the values of x and y in the third, which must also be 0, and the sign of the fourth negative for a maximum, and positive for a minimum; the discussion of the several conditions of which, although complicated, may be made in a manner similar to the above.

Examples.

1. Required to divide a number a into three parts, such that the cube of the first, into the square of the second, into the first power of the third, shall be a maximum.

Let $x =$ the first part, and $y =$ the second; then $a - x - y$ $=$ the third, and

$$u = x^3y^2(a - x - y),$$

$$\frac{du}{dx} = x^2y^2(3a - 3y - 4x), \qquad \frac{du}{dy} = x^3y(2a - 3y - 2x).$$

Placing these equal to 0, we have

$$3a - 3y - 4x = 0, \qquad 2a - 3y - 2x = 0;$$

whence $$x = \frac{a}{2}, \qquad y = \frac{a}{3}.$$

We have also

$$q = \frac{d^2u}{dx^2} = 2xy^2(3a - 3y - 6x),$$

$$q' = \frac{d^2u}{dxdy} = x^2y(6a - 9y - 8x),$$

$$q'' = \frac{d^2u}{dy^2} = x^3(2a - 6y - 2x);$$

which, for the particular values of x and y, become

$$-\frac{a^4}{9}, \qquad -\frac{a^4}{12}, \qquad -\frac{a^4}{8}.$$

Hence

$$q'^2 = \frac{a^8}{144} < qq'' = \frac{a^8}{72}, \qquad \text{and} \qquad \frac{d^2u}{dy^2} = -\frac{a^4}{8};$$

u is therefore a maximum when its value is $\dfrac{a^6}{432}$.

2. Make the preceding proposition general, by putting for the cube, square, and first power, the mth, nth, and rth powers.

Then

$$u = x^m y^n (a - x - y)^r,$$

$$x = \frac{ma}{m + n + r}, \qquad y = \frac{na}{m + n + r}.$$

3. Required the shortest distance from a given point to a given plane.

Let the equation of the plane be placed under the form

$$z = cx + dy + g,$$

and the co-ordinates of the given point be x', y', and z'; then

$$u = \sqrt{(x - x')^2 + (y - y')^2 + (z - z')^2},$$

or putting for z its value,

$$u = \sqrt{(x - x')^2 + (y - y')^2 + (cx + dy + g - z')^2}.$$

Calling the radical, R, we shall have

$$\frac{du}{dy} = \frac{y - y' + (cx + dy + g - z')d}{R},$$

$$\frac{du}{dx} = \frac{x - x' + (cx + dy + g - z')c}{R}.$$

Placing these equal to 0, and solving the resulting equations, we may obtain the values of x and y, and thence of z. Or other wise, putting for $cx + dy + g$ its value z, we have

$$y - y' + d(z - z') = 0, \quad \text{and} \quad x - x' + c(z - z') = 0,$$

which are evidently the equations of a perpendicular to the plane, and if combined with the equation of the plane will give the values of x, y, and z.

4. The volume of a rectangular parallelopipedon being given, required its three edges when its surface is a minimum.

5. Required the maximum rectangular parallelopipedon which can be inscribed in a sphere.

78. In order that a function of three or more variables be a maximum or a minimum, we must have

$$\frac{du}{dx} = 0, \qquad \frac{du}{dy} = 0, \qquad \frac{du}{dz} = 0 \ldots\ldots \&c.;$$

and the relation between the partial differential coefficients of the second order must be such, that the second term, in the development of the difference $u' - u$ shall remain of the same sign, for all the consecutive values of the function.

Geometrical Signification of a Function of a single Variable, and of its Differential Coefficient.

79. Let y be any function of x expressed by the symbol

$$y = f(x),$$

and let any value be assigned to x, and the corresponding value of y be deduced; these two values may be regarded, the first as the abscissa and the second as the ordinate of a point which may be constructed. Any number of values may thus be assigned to x, the corresponding values of y deduced, and a series of points thus constructed, through which, if a line be traced, y will be its variable ordinate and x the abscissa. Hence, we conclude that *every function of a single variable may be regarded as the ordinate of a line, of which the variable is the abscissa.*

80. Let BMM′ be a curve, the equation of which is $y = f(x)$, and M any point of this curve, the co-ordinates being x and y.

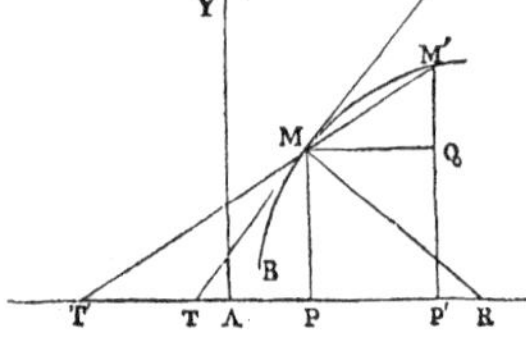

Increase the abscissa AP or x, by the variable increment PP′ $= h$; denote the corresponding ordinate P′M′ by y'; draw the secant M′MT′, the tangent MT, and MQ parallel to AX. Then

$$\text{M}'\text{Q} = \text{P}'\text{M}' - \text{PM} = y' - y = \text{P}h + \text{P}'h^2 \ldots. \text{Art. (12)}$$

From the triangle M′MQ, we have

$$\tan \text{M}'\text{MQ} = \frac{\text{M}'\text{Q}}{\text{MQ}} = \tan \text{MT}'\text{X},$$

and placing for M′Q and MQ = PP′, their values, this becomes

$$\tan \mathrm{MT'X} = \frac{\mathrm{P}h + \mathrm{P'}h^2}{h} = \mathrm{P} + \mathrm{P'}h \ldots\ldots (1).$$

Now, if h be diminished, the point M′ approaches M, and the secant M′T′ approaches the tangent MT, and finally, when $h = 0$, the point M′ coincides with M, and the secant with the tangent. If then, in (1), we make $h = 0$, we have

$$\tan \mathrm{MTX} = \mathrm{P} = \frac{dy}{dx};$$

that is, *the tangent of the angle which a tangent at any point of a line makes with the axis of X, is equal to the first differential coëfficient of the ordinate of the line.*

To show the application of this principle, let us take the equation of a circle,

$$x^2 + y^2 = \mathrm{R}^2;$$

whence

$$\frac{dy}{dx} = -\frac{x}{y} \ldots\ldots\ldots\ldots\ldots\ldots (2),$$

for the general value of the tangent of the angle, made by a tangent at any point of the circumference, with the axis of X.

If the particular value at a point whose co-ordinates are x'' and y'' be required, for x and y let x'' and y'' be substituted; then

$$\frac{dy''}{dx''} = -\frac{x''}{y''}.\text{*}$$

---

* The symbols $\frac{dy''}{dx''}$, $\frac{d^2y''}{dx''^2}$, &c., are used to indicate what the first, second, &c., differential coefficients become, when for the general variables x and y the particular values x'' and y'' are substituted; and are called the first, second, &c., differential coefficients of the ordinate of the curve taken at the point x'', y''.

Take also the equation of the conic sections,

$$y^2 = 2px + r^2x^2;$$

whence

$$\frac{dy}{dx} = \frac{p + r^2x}{y} = \frac{p + r^2x}{\sqrt{2px + r^2x^2}}.$$

For the particular point y'' and x'', this expression becomes

$$\frac{dy''}{dx''} = \frac{p + r^2x''}{\sqrt{2px'' + r^2x''^2}}.$$

81. If it be required to find the point of a given curve, at which the tangent line makes a given angle with the axis of X, we know that at this point the first differential coefficient must be equal to the tangent of the given angle. Calling this tangent a, we must then have

$$\frac{dy}{dx} = a;$$

and this, combined with the equation of the curve, will give the particular values of x and y, for the required point.

If the tangent line is to be parallel to the axis of X, then for the point of tangency, $\frac{dy}{dx} = 0$; and if perpendicular, $\frac{dy}{dx} = \infty$.

We will illustrate each of these cases by an example.

1. Let it be required to find the point on a given parabola, at which the tangent line makes an angle of 45° with the axis. The equation of the parabola is $y^2 = 2px$, by the differentiation of which, &c., we have

$$\frac{dy}{dx} = \frac{p}{y}.$$

But as $\tan 45° = 1$, we have, for the required point,

$$\frac{dy}{dx} = \frac{p}{y} = 1;$$

and, combining this with the equation $y^2 = 2px$, we fin.

$$x = \frac{p}{2}, \qquad y = p.$$

The tangent at the extremity of the ordinate passing through the focus, will then fulfil the required condition.

2. Let

$$y = a + (c - x)^2 \ldots\ldots\ldots (1)$$

represent a curve; then

$$\frac{dy}{dx} = -\,2\,(c - x),$$

which is equal to 0, when $x = c$; and this value of x in (1) gives $y = a$. These are then the co-ordinates of the point at which the tangent is parallel to the axis of X.

3. Let

$$y = a + (c - x)^{\frac{1}{2}}$$

represent a curve; then

$$\frac{dy}{dx} = -\,\frac{1}{2\,(c - x)^{\frac{1}{2}}},$$

which is equal to infinity, when $x = c$. $x = c$ and $y = a$ are then the co-ordinates of the point at which the tangent is perpendicular to the axis of X.

Equations of Tangent and Normal. Expressions for Sub-Tangent, Sub-Normal, &c.

82. If x'' and y'' represent the co-ordinates of a given point on a given curve, whose equation is $y = f(x)$, the equation of a straight line passing through this point will be

$$y - y'' = a(x - x'') \dots\dots (1),$$

a being indeterminate. In order that this line be a tangent at the given point, a must be the first differential coefficient of the ordinate of the curve taken at this point, Note, Art. (80); that is, for a we must substitute $\frac{dy''}{dx''}$. We thus obtain

$$y - y'' = \frac{dy''}{dx''}(x - x'') \dots\dots (2).$$

By differentiating the equation of an ellipse,

$$a^2y^2 + b^2x^2 = a^2b^2,$$

we deduce

$$\frac{dy}{dx} = -\frac{b^2x}{a^2y}; \qquad \text{whence} \qquad \frac{dy''}{dx''} = -\frac{b^2x''}{a^2y''};$$

and this value in (2) gives, for the equation of a tangent to an ellipse at the point y'', x'',

$$y - y'' = -\frac{b^2x''}{a^2y''}(x - x''),$$

which, by reduction, becomes $a^2yy'' + b^2xx'' = a^2b^2$.

83. If the equation of a tangent be required, which shall be parallel to a given line, or make a given angle with the axis of X, we may determine the co-ordinates of the point of contact as in article (81); and knowing these, the equation may be deduced as above.

Thus, if a tangent to a circle be required to make with the axis of X an angle whose tangent is 2, we must have for the required point, equation (2), Art. (80),

$$\frac{dy}{dx} = -\frac{x}{y} = 2.$$

From this, we find $y = -\frac{x}{2}$, which, combined with the equation of the circle, gives

$$x = \pm\frac{2\,\text{R}}{\sqrt{5}} = x'', \qquad y = \mp\frac{\text{R}}{\sqrt{5}} = y'';$$

and equation (2), Art. (82), becomes, when we use the upper signs,

$$y + \frac{\text{R}}{\sqrt{5}} = 2\left(x - \frac{2\,\text{R}}{\sqrt{5}}\right), \quad \text{or} \quad y = 2x - \text{R}\sqrt{5}.$$

84. Equation (1), Art. (82), will become the equation of a normal at the point x'', y'', if for a, we substitute

$$-\frac{1}{\frac{dy''}{dx''}} = -\frac{dx''}{dy''} \ldots .(\text{Analyt. Geo.});$$

and we thus obtain for the general equation of a normal,

$$y - y'' = -\frac{dx''}{dy''}(x - x'').$$

85. The right-angled triangle MTP (Figure of Art. 80) gives

$$\mathrm{PM} = \mathrm{PT} \tan \mathrm{MTP}; \qquad \text{hence} \qquad \mathrm{PT} = \frac{\mathrm{PM}}{\tan \mathrm{MTP}};$$

or

$$\textit{Subtangent} = \frac{y}{\frac{dy}{dx}} = y\frac{dx}{dy}.$$

Also,

$$\mathrm{MT} = \sqrt{\overline{\mathrm{MP}}^2 + \overline{\mathrm{PT}}^2},$$

or

$$\textit{Tangent} = \sqrt{y^2 + y^2\frac{dx^2}{dy^2}} = y\sqrt{1 + \frac{dx^2}{dy^2}}.$$

The right-angled triangle PMR gives

$$\mathrm{PR} = \mathrm{MP} \tan \mathrm{PMR}; \qquad \text{but} \qquad \mathrm{PMR} = \mathrm{MTP};$$

hence,

$$\mathrm{PR} = \mathrm{MP} \tan \mathrm{MTP}, \qquad \text{or} \qquad \textit{Subnormal} = y\frac{dy}{dx}.$$

Also,

$$\mathrm{MR} = \sqrt{\overline{\mathrm{MP}}^2 + \overline{\mathrm{PR}}^2};$$

hence,

$$\textit{Normal} = \sqrt{y^2 + y^2\frac{dy^2}{dx^2}} = y\sqrt{1 + \frac{dy^2}{dx^2}}.$$

To apply these formulas to a particular curve, it is only necessary to substitute in each the expression for $\frac{dx}{dy}$, or $\frac{dy}{dx}$, deduced from the differential equation of the curve. The results will be general for all points of the curve. If the values for a given point be required, in these results let the co-ordinates of the point be substituted for x and y.

For example, take the general equation of Conic Sections,

$$y^2 = 2px + r^2x^2;$$

whence

$$\frac{dy}{dx} = \frac{p + r^2x}{\sqrt{2px + r^2x^2}}, \qquad \frac{dx}{dy} = \frac{\sqrt{2px + r^2x^2}}{p + r^2x}.$$

These expressions substituted in the formulas, give

$$\text{PT} = \frac{2px + r^2x^2}{p + r^2x}, \qquad \text{PR} = p + r^2x,$$

$$\text{MT} = \sqrt{2px + r^2x^2 + \left(\frac{2px + r^2x^2}{p + r^2x}\right)^2},$$

$$\text{MR} = \sqrt{2px + r^2x^2 + (p + r^2x)^2}.$$

For the parabola $r^2 = 0$, and these expressions become,

$$\text{PT} = 2x, \qquad \text{PR} = p,$$

$$\text{MT} = \sqrt{2px + 4x^2}, \qquad \text{MR} = \sqrt{2px + p^2}.$$

Convexity and Concavity of Curves.

86. A curve, at a point, is *convex* towards another line, when, in the immediate vicinity of the point, its tangent lies between it and the line. It is *concave* when it lies between its tangent and the line.

If a curve be convex towards the axis of X, and the ordinate *positive*, as in the annexed figure, it is plain, that as the abscissas AP, AP′, &c., increase, the angles MTX, M′T′X, &c., will increase, and the reverse; consequently their tangents will also increase as x increases, or decrease as x decreases. Since these tangents are represented by the corresponding values of the first differential coefficient of the ordinate $\left(\frac{dy}{dx}\right)$, *it must be an increasing function of x*, and its differential coefficient, i. e., $\frac{d^2y}{dx^2}$, must be *positive*, Art. (14).

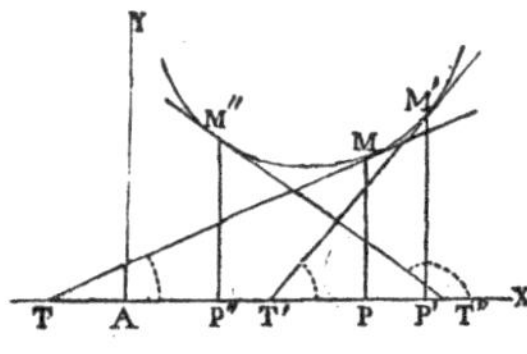

If the curve be still convex, and the ordinate *negative*, the angles STX, S′T′X, &c., and their tangents plainly decrease as x increases; $\frac{dy}{dx}$ is a decreasing function of x, and $\frac{d^2y}{dx^2}$ must be *negative*.

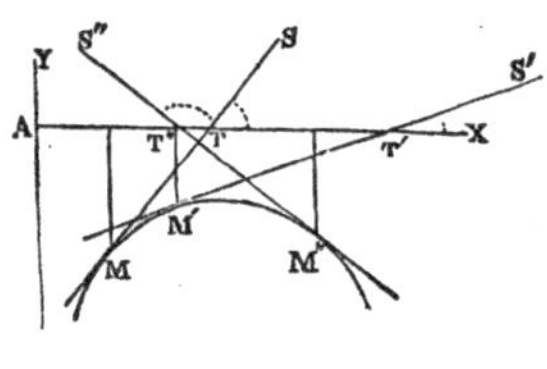

If, then, a curve be convex towards the axis of abscissas, *the ordinate and its second differential coefficient, taken at the different points, will have the same sign.*

If the curve be concave, and the ordinate *positive*, as in the figure, the angles MTX, M′T′X, &c., and their tangents will decrease as x increases; $\frac{dy}{dx}$ will be a decreasing function, and $\frac{d^2y}{dx^2}$ *negative.*

If the curve be concave, and the ordinate *negative*, the reverse will evidently be the case, and $\frac{d^2y}{dx^2}$ will be *positive*.

Hence, if a curve be concave towards the axis of abscissas, *he ordinate and its second differential coefficient will have contrary signs.*

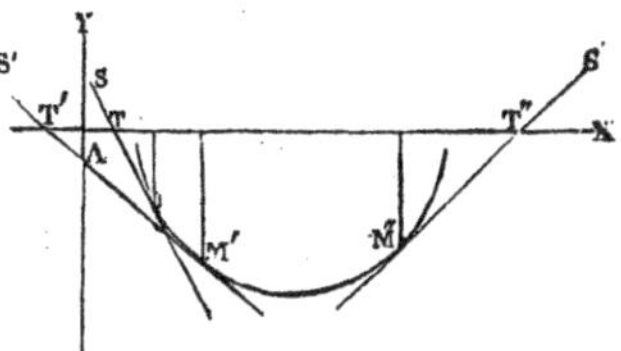

ASYMPTOTES.

87. An asymptote is a line which, continually approaching a curve, becomes tangent to it at an infinite distance. Asymptotes may be *curvilinear* or *rectilinear*. The latter only will be considered here.

Let MV be any curve, and BC a rectilinear asymptote. Also, let MT be a tangent, the equation of which, article (82), is

$$y - y'' = \frac{dy''}{dx''}(x - x'').$$

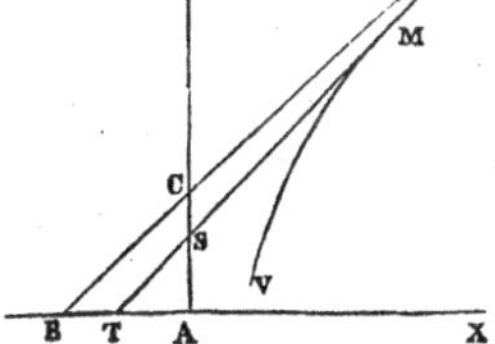

If we make $y = 0$ in this equation, we obtain

$$x = x'' - y''\frac{dx''}{dy''} = \text{AT}\ldots\ldots(1).$$

If we make $x = 0$, we obtain

$$y = y'' - x''\frac{dy''}{dx''} = \text{AS}\ldots\ldots(2).$$

Now, as the point of contact, M, the co-ordinates of which are x'' and y'', is removed farther from the origin, the tangent MT will approach nearer to the asymptote BC; and finally, when M is at an infinite distance, the two will coincide, and the distances AT and AS become respectively equal to the distances AB and AC.

If, then, the expressions for AT and AS, *when such values are substituted for x'' and y'' as to remove the point M to an infinite distance*, are both finite, there will be an asymptote, which may be drawn through the points B and C.

If one of these expressions becomes infinite, and the other finite, there will be an asymptote parallel to the axis, on which the distance is infinite.

If both expressions become infinite, there will be no asymptote. If both become 0, the asymptote will pass through the origin of co-ordinates, and the tangent of the angle which it makes with the axis of X may be obtained from the expression for $\frac{dy''}{dx''}$, when for x'' and y'' the proper values are substituted.

Hence, to construct the asymptote of a given curve: Find, by differentiating the equation of the curve, the expressions for $\frac{dy''}{dx''}$ and $\frac{dx''}{dy''}$, which substitute in formulas (1) and (2); the results thence obtained by substituting for x'' and y'' their values for that point of the curve which is at an infinite distance, will be the distances cut off from the co-ordinate axes by the asymptote, if there is one.

Examples.

1. Take the equation of lines of the second order,

$$y^2 = 2px + r^2x^2.$$

By differentiation, &c., we obtain

$$\frac{dy}{dx} = \frac{p + r^2x}{y} = \frac{p + r^2x}{\pm\sqrt{2px + r^2x^2}};$$

whence

$$\frac{dy''}{dx''} = \frac{p + r^2x''}{\pm\sqrt{2px'' + r^2x''^2}}.$$

Substituting this and the value of $y'' = \pm\sqrt{2px'' + r^2x''^2}$ in (1) and (2), we have

$$AT = x'' - \frac{2px'' + r^2x''^2}{p + r^2x''} = \frac{-px''}{p + r^2x''} = \frac{-p}{\frac{p}{x''} + r^2} \ldots\ldots(3).$$

$$AS = \pm\sqrt{2px''+r^2x''^2} - \frac{px''+r^2x''^2}{\pm\sqrt{2px''+r^2x''^2}} = \frac{p}{\pm\sqrt{\frac{2p}{x''}+r^2}} \ldots(4).$$

In this case, the co-ordinates of that point of the curve which is at an infinite distance are $x'' = \infty$ and $y'' = \infty$. Making $x'' = \infty$ in (3) and (4), we have

$$AB = -\frac{p}{\frac{p}{\infty} + r^2} = -\frac{p}{r^2}.$$

$$AC = \frac{p}{\pm\sqrt{\frac{2p}{\infty} + r^2}} = \pm\frac{p}{\sqrt{r^2}}.$$

For the hyperbola r^2 is positive, these expressions are both finite, and, as there are two different values of AC, there are two asymptotes; and since $p = \frac{b^2}{a}$ and $r^2 = \frac{b^2}{a^2}$, we have

$$AB = -a, \qquad AC = \pm b.$$

For the parabola $r^2 = 0$, the expressions are both infinite, and there is no asymptote.

For the ellipse r^2 is negative, and AC imaginary, as it should be, since there is no point of the curve at an infinite distance, and of course no asymptote.

2. Take the equation

$$x^3 - 3axy + y^3 = 0.$$

From formulas (1) and (2), we obtain in this case

$$\text{AT} = \frac{ax''y''}{x''^2 - ay''} \dots\dots (5), \qquad \text{AS} = \frac{ax''y''}{y''^2 - ax''} \dots\dots (6).$$

As it is difficult to obtain the value of y in terms of x from the given equation, we cannot at once eliminate y'' from (5) and (6); but if we make $x = ty$ and substitute in the given equation, it will be divisible by y^2, and we then find

$$y = \frac{3at}{1 + t^3}.$$

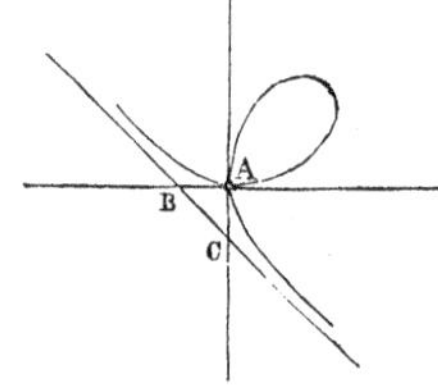

If in this we make $t = -1$, we have $y = \infty$; hence, when y is infinite it is equal to $-x$, and for that point of the curve which is at an infinite distance we have $y'' = -x'' = \infty$.

Changing y'' into $-x''$ in (5) and (6), they become

$$\text{AT} = \frac{-a}{1 + \frac{a}{x''}}, \qquad \text{AS} = \frac{-a}{1 - \frac{a}{x''}};$$

and making $x'' = \infty$, we find

$$AB = -a, \qquad \text{and} \qquad AC = -a;$$

hence, BC is the asymptote.

3. Take the equation

$$x^2y^3 = p,$$

in the curve represented by which, the points at an infinite distance have for their co-ordinates

$$x'' = 0, \quad y'' = \infty, \qquad \text{and} \qquad y'' = 0, \quad x'' = \infty.$$

Advantages of regarding the Differential of the Independent Variable as infinitely small.

88. Heretofore, in our treatment of the subject, we have regarded the differential of the independent variable merely as a constant, Art. (9), without having fixed upon the particular value for it. All the demonstrations are then as true for one value as for another.

It is, however, of the greatest convenience, in the application of the Calculus to the higher branches of Mathematics and Physical Science, to regard this differential as *infinitely small*, that is, *so small as to be contained in unity an infinite number of times;* and hereafter it will be so regarded.

The advantages of so regarding it will appear evident after a few illustrations. Let us take first the simple function discussed in article (9),

$$u = ax^2.$$

After x has been increased by dx, we have there found

$$u' - u = 2axdx + adx^2.$$

Now, if the increment (dx) of the variable be infinitely small, the two states u and u' will plainly be *consecutive*, the expression for their difference being

$$2axdx + adx^2 \ldots\ldots\ldots\ldots (1).$$

But since dx is infinitely small, its square will be infinitely small when compared with it; as may be shown by taking the identical equation

$$\frac{1}{dx} = \frac{dx}{dx^2} = \frac{dx^2}{dx^3} = \&c.;$$

from which, since dx is contained an infinite number of times in unity, it appears that dx^2 will be contained an infinite number of times in dx; dx^3 in dx^2, &c.: adx^2 will then be infinitely small with reference to $2axdx$, and its addition will give no sensible increase; hence, adx^2 may be omitted from expression (1) without materially affecting its value, and in this case $2axdx$ *may be taken for, or is the measure of, the difference* $u' - u$.

This is true also in the general case; for if we take equation (3), Art. (9), after substituting dx for h, we have

$$u' - u = \text{P}dx + \text{Q}dx^2 + \text{R}dx^3 + \&c.;$$

in which all the terms except $\text{P}dx$, which is the differential of the function, contain the second or a higher power of dx, and are therefore infinitely small with reference to this first term; and may be rejected, and the differential of the function taken, as *the measure of the difference between two consecutive states of the function.*

It is plain, also, since

$$du = pdx, \quad d^2u = qdx^2, \quad d^3u = rdx^3, \ \&c. \ldots \text{Art. (28)},$$

that the second differential of a function is infinitely small when compared with the first, and the third when compared with the second, and so on. It is usual to call these, infinitely small quantities of the first, second, and third orders; and we see, from what precedes, that *every infinitely small quantity may be omitted without error, when connected by the sign $\pm$ with a finite quantity, or with an infinitely small quantity of a lower order.*

In the application of the Calculus to curves, these principles are of great use. Let BMM′ be a curve; MP, M′P′, any two consecutive ordinates; PP′ = P′P″ = P″P‴, &c., being each equal to dx; then the difference between y and y', or $y' - y = $ M′Q, is equal to dy; and $z' - z$ = MM′ $= dz$. Or, since $z' - z$ may represent the difference MM′, M′M″, M″M‴, between any two consecutive states of the arc, the different values of dz may, in succession, represent the infinitely small elementary arcs MM′, M′M″, &c., the sum of all of which will be equal to the entire arc z.

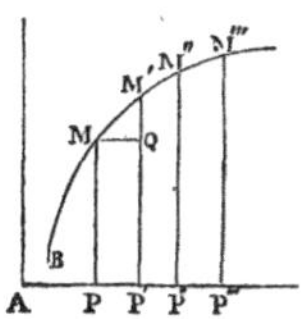

So the difference between the two areas BM′P′ and BMP is equal to PMM′P′ $= ds$; and the different values of ds may, in succession, represent the infinitely small elementary areas PMM′P′, P′M′M″P″, &c., the sum of all of which will equal the entire area s. And in general, if the variable be increased by its differential, *the corresponding increment of the function may be represented by the differential of the function; and the sum of all the different values of this differential will equal the function itself. The differential coefficient will, in this case, express the rate of increase or change of the function, in passing from state to state.*

DIFFERENTIALS OF AN ARC, PLANE AREA, SURFACE AND VOLUME OF REVOLUTION.

89. The differentials of an arc, plane area, &c., which, as will be seen, are functions of a single variable, may be obtained by the application of the general rule in Art. (9); but as the process in these cases is long and complicated, it will be found much more simple to make use of the principles developed in the preceding article.

90. Let BM $= z$ be any arc of a curve, the equation of which is $y = f(x)$. Although z changes whenever x or y is changed, yet the equation $y = f(x)$ establishes such a relation between x and y that one is necessarily a function of the other. z may, therefore, be regarded as a function of either. Let us regard it as a function of x, and let AP $= x$, PM $= y$, and increase x by PP′ $= dx$; then, Art. (88),

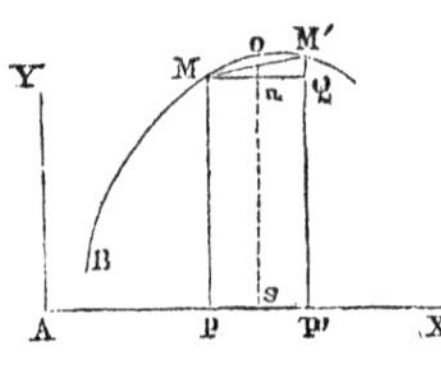

$$\text{BM}' = z', \quad \text{MM}' = z' - z = dz, \quad \text{M}'\text{Q} = y' - y = dy.$$

Since the arc MM′ is infinitely small, it will not differ from its chord MM′, which is therefore equal to dz. From the right-angled triangle MQM′, we have

$$\overline{\text{MM}'} = \sqrt{\overline{\text{MQ}}^2 + \overline{\text{M}'\text{Q}}^2}, \qquad \text{or} \qquad dz = \sqrt{dx^2 + dy^2};$$

that is, *the differential of an arc is equal to the square root of the sum of the squares of the differentials of the co-ordinates of its points.*

Since we regard z as a function of x, we should express its differential in terms of x alone. To do this, we differentiate the given equation of the curve, combine the differential equation thus deduced with the given equation, and obtain an expression for dy in terms of x and dx, and substitute in the above formula. If z should be regarded as a function of y, we should, by the same process, obtain an expression for dx in terms of y and dy, and substitute in the formula.

To illustrate, take the equation of a circle,

$$x^2 + y^2 = \text{R}^2;$$

whence

$$dy = -\frac{xdx}{y} = -\frac{xdx}{\sqrt{\text{R}^2 - x^2}},$$

and

$$dz = \sqrt{dx^2 + \frac{x^2dx^2}{\text{R}^2 - x^2}} = \frac{\text{R}dx}{\sqrt{\text{R}^2 - x^2}}.$$

91. If x, y, and z denote the co-ordinates of the points of a curve s in space, ds will be the diagonal of a rectangular parallelopipedon, and we deduce

$$ds = \sqrt{dx^2 + dy^2 + dz^2}.$$

92. Let $BMP = s$ be any plane area, limited by a curve and the axis of X; it will evidently be a function of x.

Increase x by $PP' = dx$; then, in the figure of Art. (90),

$$PMM'P' = s' - s = ds;$$

but since the chord MM′ coincides with the arc MM′, this area is the same as that of the trapezoid PMM′P′, the measure of which, o being the middle point of MM′, is $os \times PP'$. But

$$os = MP + no = y + \tfrac{1}{2}dy;$$

hence,

$$PMM'P' = (y + \tfrac{1}{2}dy)\,dx,$$

or, rejecting $\frac{1}{2}dy$,

$$ds = y\,dx \ldots\ldots\ldots\ldots (1);$$

that is, *the differential of the area is equal to the ordinate of the bounding curve into the differential of the abscissa.*

The differential of the area included between the curve and axis of Y, may be found in the same way to be

$$ds = xdy \ldots\ldots\ldots\ldots (2).$$

If the axes of co-ordinates are oblique to each other, PMM′P′ would be a trapezoid, and its area would be $y\,dx \sin\beta$, β being the angle made by the co-ordinate axes; hence, in this case,

$$ds = y \sin\beta\,dx \ldots\ldots\ldots (3).$$

To express these differentials in terms of a single variable, we find, from the equation of the curve, the value of y in terms of x; or from this equation and its differential equation, deduce the value of dx in terms of y and dy, and substitute in equations (1) and (3); or deduce in a similar way x or dy, and substitute in equation (2).

For examples, take the equation

$$y^2 = R^2 - x^2;$$

whence

$$ds = ydx = \sqrt{R^2 - x^2}dx.$$

Also the equation

$$y^2 = 2p'x,$$

the axes of co-ordinates being oblique; then

$$ds = \sqrt{2p'x} \sin \beta dx.$$

93. Let the curve BM revolve about the axis of X; it will generate a surface of revolution which will be a function of x, and which we denote by u.

The notation being as in the preceding articles, the increment $(u' - u)$ of the surface, when x is increased by dx, will be generated by the arc MM'. But this surface is the same as that of the frustum of the cone, generated by the chord MM', which is measured by circumference os multiplied by MM'; whence

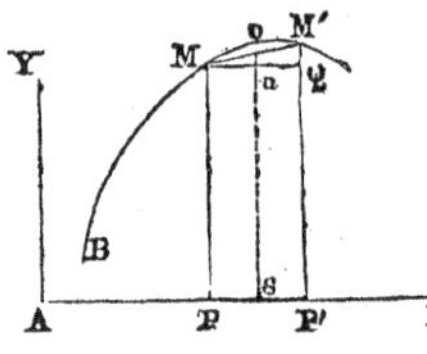

$$du = 2\pi(y + \tfrac{1}{2}dy)\, dz = 2\pi y\, dz,$$

rejecting $\frac{1}{2}dy$. Substituting for dz its value, Art. (90), we have

$$du = 2\pi y \sqrt{dx^2 + dy^2} \ldots\ldots (1);$$

that is, the differential of a surface of revolution *is equal to the circumference of a circle perpendicular to the axis, multiplied by the differential of the arc of the generating curve.* If the curve revolve about the axis of Y, we may determine in the same way

$$du = 2\pi x \sqrt{dx^2 + dy^2}.$$

To obtain the differential in terms of a single variable, we find from the equation and differential equation of the generating curve, the expressions for y and dy in terms of x, or of dx in terms of y and dy, and substitute in (1).

If we suppose a parabola, whose equation is $y^2 = 2px$, to revolve about its axis, we shall have

$$dx = \frac{ydy}{p},$$

$$du = 2\pi y \sqrt{\frac{y^2dy^2}{p^2} + dy^2} = \frac{2\pi ydy}{p}\sqrt{y^2 + p^2}.$$

94. Let the area BMP revolve about the axis of X; it will generate a volume of revolution, which is a function of x, and which we denote by v. If x be increased by $PP' = dx$, then the area PMM′P′ will generate the increment $(v' - v)$ of the volume. But this is the same as that of the frustum generated by the trapezoid PMM′P′, which is measured by

$$\tfrac{1}{3}[y^2 + (y + dy)^2 + y(y + dy)]\,dx;$$

hence, rejecting dy, we have

$$dv = \pi y^2 dx \ldots\ldots\ldots.(1);$$

that is, the differential of a volume of revolution *is equal to the area of a circle perpendicular to the axis, multiplied by the differential of the abscissa of the curve which generates the bounding surface.*

For the volume generated by the area included between the curve and axis of Y, we may find, in the same way,

$$dv = \pi x^2 dy.$$

By deducing the value of y^2 in terms of x from the equation of the meridian or generating curve, or the value of dx in terms of y and dy from this equation and its differential equation, and substituting in (1), we shall have the differential of the volume in terms of one variable and its differential.

If we take the particular case of the ellipsoid, the equation of the generating curve being

$$y^2 = \frac{b^2}{a^2}(a^2 - x^2),$$

we have

$$dv = \pi y^2 dx = \pi \frac{b^2}{a^2}(a^2 - x^2)\,dx.$$

Tendency of Curves to coincide. Osculatory Curves and Curvature.

95. It is now proposed to examine the tendency which curves, with a common point, have to coincide with each other in the vicinity of this point; and also the use which may be made of this property of curves. Let

$$y = f(x), \qquad \text{and} \qquad y' = f'(x'),$$

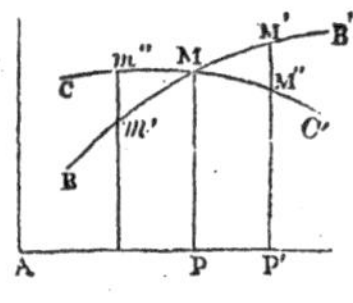

be the equations of any two curves BB′ and CC′, having the point M in common. Increase the abscissa of each by h; then, by Art. (35), we have for the second states of the ordinates,

$$f(x+h) = y + \frac{dy}{dx}h + \frac{d^2y}{dx^2}\frac{h^2}{1.2} + \frac{d^3y}{dx^3}\frac{h^3}{1.2.3} + \&c.\dots(1),$$

$$f'(x'+h) = y' + \frac{dy'}{dx'}h + \frac{d^2y'}{dx'^2}\frac{h^2}{1.2} + \frac{d^3y'}{dx'^3}\frac{h^3}{1.2.3} + \&c.\dots(2).$$

As these curves have the point M in common, for this point we must have

$$x = x', \qquad \text{and} \qquad y = y';$$

and if the co-ordinates of this point be substituted in the preceding equations, $f(x + h)$ and $f'(x' + h)$ will represent, respectively, the ordinates M′P′ and M″P′; $\frac{dy}{dx}$, $\frac{d^2y}{dx^2}$, &c., will represent the successive differential coefficients of the ordinate of the first curve *taken at the point* M, Note, Art. (80); and $\frac{dy'}{dx'}$, $\frac{d^2y'}{dx'^2}$, &c., corresponding values for the second curve.

If, under this supposition, we subtract equation (2) from (1), member from member, and place

$$\frac{dy}{dx} - \frac{dy'}{dx'} = A', \qquad \frac{d^2y}{dx'^2} - \frac{d^2y'}{dx'^2} = A'', \quad \&c.,$$

we obtain

$$M'M'' = A'h + A''\frac{h^2}{1.2} + A'''\frac{h^3}{1.2.3} + \&c.\dots(3).$$

If now we make h infinitely small, each term of the series will exceed the sum of all the succeeding terms, Art. (13), the points M′ and M″ will be consecutive with M, and it is evident that the smaller the distance M′M″ becomes, the greater will be the tendency of the curve CC′ to coincidence with BB′, in the vicinity of M. If $A' = 0$, or

$$\frac{dy}{dx} = \frac{dy'}{dx'},$$

by which M'M'' will be made very small, the curves are said to have a contact *of the first order*. In this case they have a common tangent at M, or are simply tangent, Art. (80).

If A'' = 0, also, or

$$\frac{d^2y}{dx^2} = \frac{d^2y'}{dx'^2},$$

the second member of equation (3) becomes still less, the tendency to coincidence still greater, and the curves are said to have a contact *of the second order*.

If A''' = 0, the second member is smaller still, and the condition

$$\frac{d^3y}{dx^3} = \frac{d^3y'}{dx'^3}$$

gives the curves a contact *of the third order*; and, in general, two curves have a contact *of the mth order* when they have a point in common, and the first m differential coefficients of their ordinates, taken at this point, respectively equal.

To ascertain the order of contact of two given curves, we have simply to combine their equations and deduce the values of the variables; for each set of real values they will have a common point; then differentiate both equations, deduce the first differential coefficient of the ordinate of each curve, and substitute in them the co-ordinates of the common point; if the results are equal, the curves have a contact of the first order. Differentiate again, and find the second differential coefficients of the ordinates taken at this point; if they are equal, the curves have a contact of the second order; and so on, the order of contact being always denoted by the number of equal successive differential coefficients thus taken.

To illustrate, take the two equations

$$y^2 = 4x \ldots\ldots\ldots (1), \qquad y = x + 1 \ldots\ldots\ldots (2).$$

By combining them we find a common point, the co-ordinates of which are

$$x'' = 1, \qquad y'' = 2.$$

By differentiation, we find from (1),

$$\frac{dy}{dx} = \frac{2}{y} \ldots\ldots\ldots (3), \qquad \text{whence} \qquad \frac{dy''}{dx''} = 1;$$

and from (2),

$$\frac{dy}{dx} = 1 \ldots\ldots\ldots (4), \qquad \text{whence} \qquad \frac{dy''}{dx''} = 1.$$

Differentiating again, we have from (3),

$$\frac{d^2y}{dx^2} = -\frac{4}{y^3}, \qquad \text{whence} \qquad \frac{d^2y''}{dx''^2} = -\frac{1}{2};$$

and from (4),

$$\frac{d^2y}{dx^2} = 0, \qquad \text{whence} \qquad \frac{d^2y''}{dx''^2} = 0.$$

The two lines having a point in common, and the first differential coefficients of the ordinate taken at this point equal, have a contact of the first order. Since the second differential coefficients are not equal, the order of contact is no higher than the first.

Take also the equations,

$$4y = x^2 - 4, \qquad y^2 - 2y = 3 - x^2,$$

and ascertain the order of contact of the curves.

96. The constants which enter into the equation of a curve determine its extent and position with respect to the co-ordinate axes. If, then, one curve be given completely, and another in kind only, by its general equation, the constants in this equation being arbitrary, we can evidently assign such values to them as shall require this curve to fulfil as many conditions as its equation contains constants; that is, we may cause its equation to be satisfied by the substitution of the co-ordinates of a given point of the first curve, thus making it pass through this point; and, in addition, may make as many differential coefficients of its ordinate taken at this point, equal to the corresponding ones of the first, as there are constants to be disposed of, less one, thus giving to the second curve an order of contact at a given point of the first, *denoted by the number of constants less one.*

To ascertain the values which must be assigned to the arbitrary constants, first substitute the co-ordinates of the given point in the equation of the second curve; obtain then the first differential coefficients of the ordinate by differentiating the equation of each curve, substitute in these the co-ordinates of the given point, and place the results equal; do the same with the successive differential coefficients, until as many equations are thus formed as there are arbitrary constants. By the solution of these equations, we can find those values of the constants which will cause the conditions to be fulfilled. These, substituted in the equation of the second curve, will give an equation which will represent the particular curve having the required order of contact.

The curve which, at a given point of a given curve, has a higher order of contact than any other of the same kind, is called *an osculatrix.* Thus, *an osculatory circle is one which has a higher order of contact than any other circle.*

Since no more conditions can be assigned than there are constants, *the highest order of contact which can be given to a curve is denoted by the number of constants, less one, which enter its most general equation.*

97. Let these principles be applied:

First: *To find the equation of an osculatory right line.*

Let the equation of the given curve be

$$y = f(x),$$

and the co-ordinates of the given point, x'' and y''.

The most general equation of the right line is

$$y = ax + b \dots\dots\dots\dots (1),$$

containing but two arbitrary constants. The first condition to be fulfilled is that the co-ordinates x'' and y'' shall satisfy equation (1); that is, we must have

$$y'' = ax'' + b \dots\dots\dots (2).$$

The first differential coefficient of the ordinate derived from the equation of the given curve is $\frac{dy}{dx}$, which for the given point becomes $\frac{dy''}{dx''}$. The first differential coefficient derived from equation (1) is $\frac{dy}{dx} = a$; hence, the second condition is

$$\frac{dy''}{dx''} = a \dots\dots\dots\dots (3).$$

By the solution of equations (2) and (3), we find

$$a = \frac{dy''}{dx''}, \qquad b = y'' - \frac{dy''}{dx''}x''.$$

These values in (1), give the equation

$$y = \frac{dy''}{dx''}x + y'' - \frac{dy''}{dx''}x'', \quad \text{or} \quad y - y'' = \frac{dy''}{dx''}(x - x'').$$

This, as it should be, is the same equation as that deduced in Art. (82).

98. Second: *To find the equation of the osculatory circle at any point of the curve whose equation is* $y = f(x)$.

Denote the given point, *or point of osculation*, by x'' and y''.

The most general equation of the circle is

$$(x - \alpha)^2 + (y - \beta)^2 = R^2 \ldots\ldots (1),$$

containing three arbitrary constants. A contact of the second order may therefore be given to the circle.

Substituting x'' and y'' for x and y in the above equation, we have, for the first equation of condition,

$$(x'' - \alpha)^2 + (y'' - \beta)^2 = R^2 \ldots\ldots (2).$$

By differentiating the equation $y = f(x)$, and substituting x'' and y'' in the first and second differential coefficients, we obtain

$$\frac{dy''}{dx''}, \qquad \text{and} \qquad \frac{d^2y''}{dx''^2} \ldots\ldots (3).$$

Differentiating equation (1) twice, we have

$$(x - \alpha)\,dx + (y - \beta)\,dy = 0, \quad \text{whence} \quad \frac{dy}{dx} = -\frac{x - \alpha}{y - \beta};$$

$$dx^2 + dy^2 + (y - \beta)\,d^2y = 0, \quad \text{whence} \quad \frac{d^2y}{dx^2} = -\frac{1 + \dfrac{dy^2}{dx^2}}{y - \beta}.$$

Substituting x'' and y'' for x and y in the expressions for these differential coefficients, we obtain

$$-\frac{x'' - \alpha}{y'' - \beta}, \qquad \text{and} \qquad -\frac{1 + \dfrac{dy''^2}{dx''^2}}{y'' - \beta};$$

and placing these respectively equal to expressions (3), we have, for the second and third equations of condition,

$$\frac{dy''}{dx''} = -\frac{x'' - \alpha}{y'' - \beta} \dots (4), \qquad \frac{d^2y''}{dx''^2} = -\frac{1 + \frac{dy''^2}{dx''^2}}{y'' - \beta} \dots (5).$$

By the solution of these equations, we can find the values of R, α, and β, which, substituted in (1), will give the equation of the osculatory circle.

To illustrate, let us seek the equation of the circle osculatory to the parabola whose equation is

$$y^2 = 4x,$$

at the point whose co-ordinates are $\quad x'' = 1, \quad y'' = 2.$

Differentiating the given equation twice, and substituting the co-ordinates 1 and 2, we find

$$\frac{dy}{dx} = \frac{2}{y}, \qquad \text{whence} \qquad \frac{dy''}{dx''} = 1;$$

$$\frac{d^2y}{dx^2} = -\frac{4}{y^3}, \qquad \text{"} \qquad \frac{d^2y''}{dx''^2} = -\frac{1}{2}.$$

These values, with the co-ordinates of the given point, placed in the equations of condition, give

$$(1 - \alpha)^2 + (2 - \beta)^2 = R^2,$$

$$1 = -\frac{1 - \alpha}{2 - \beta}, \qquad -\frac{1}{2} = -\frac{2}{2 - \beta};$$

whence

$$\alpha = 5, \qquad \beta = -2, \qquad R = \sqrt{32};$$

and the equation of the osculatory circle will then be

$$(x - 5)^2 + (y + 2)^2 = 32.$$

Find also the equation of the circle osculatory to the curve represented by the equation

$$4y = x^2 - 4,$$

at the point whose co-ordinates are $x'' = 0, \quad y'' = -1$.

99. Since in the three equations of condition just considered, which are called *the equations of condition for the osculatory circle*, x'' and y'' may, in succession, be made to represent every point of the given curve, we may omit the dashes and write the equation thus,

$$(x - \alpha)^2 + (y - \beta)^2 = R^2 \ldots\ldots (1),$$

$$x - \alpha = -\frac{dy}{dx}(y - \beta) \ldots\ldots\ldots (2),$$

$$y - \beta = -\frac{dx^2 + dy^2}{d^2y} \ldots\ldots\ldots (3);$$

in which, it must be recollected, x and y are the co-ordinates of the point of osculation, α and β the co-ordinates of the centre of the osculatory circle, and R its radius.

Substituting in (1) the value of $x - \alpha$, and reducing, we obtain

$$R^2 = (y - \beta)^2 + \frac{dy^2}{dx^2}(y - \beta)^2 = (y - \beta)^2\left(\frac{dx^2 + dy^2}{dx^2}\right);$$

whence, by the substitution of the value of $y - \beta$,

$$R = \pm\frac{(dx^2 + dy^2)^{\frac{3}{2}}}{dxd^2y},$$

which is a general value for the radius of the osculatory circle.

If z denote the arc of the given curve, then

$$dz = \sqrt{dx^2 + dy^2} \ldots \text{Art. (90)};$$

hence, the above expression for R becomes

$$R = \pm \frac{dz^3}{dx\, d^2y}.$$

100. If φ denote the angle made by the radius of the osculatory circle drawn to the point of osculation, with a fixed line as OP, M and M′ two consecutive points, and MC and M′C the corresponding radii intersecting at C, then

$$MC = R, \quad MM' = dz, \quad nn' = d\varphi;$$

and we have

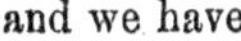

$$nC : MC :: nn' : MM',$$

or

$$1 : R :: d\varphi : dz;$$

hence,

$$dz = R d\varphi, \qquad \text{and} \qquad R = \frac{dz}{d\varphi}.$$

101. Let A be any point in the plane of an osculatory circle, C its centre, MP a tangent at the point of osculation, and AP a perpendicular to the tangent. Join AC, and denote CM by R, AM by r, and AP by p; then

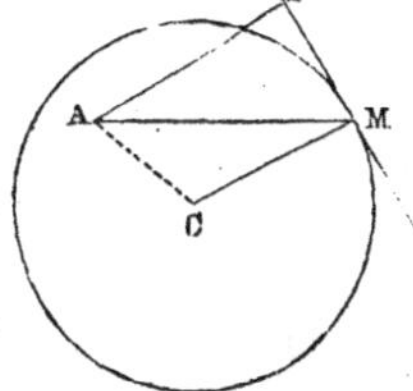

$$\overline{AC}^2 = R^2 + r^2 - 2rR \cos AMC;$$

but $$\cos \text{AMC} = \sin \text{AMP} = \frac{p}{r};$$

hence, $$\overline{\text{AC}}^2 = \text{R}^2 + r^2 - 2\text{R}p.$$

Differentiating, since AC and R are constant as we pass from one point of the circle to another, we have

$$0 = 2rdr - 2\text{R}dp,$$

or

$$\text{R} = \frac{rdr}{dp}.$$

102. If two lines have the first order of contact, or are simply tangent to each other, they are said to have two consecutive points in common. If they have the second order of contact, they have three consecutive points in common, and so on.

If, then, M, M′, and M″ be any three consecutive points of a curve of double curvature referred to the axes AX, AY, and AZ, the circle passing through these points, having its centre at C, will be the osculatory circle of the curve. It is now proposed to determine a general expression for its radius. Denote the co-ordinates of M by x, y, and z; the co-ordinates of M′ will then be

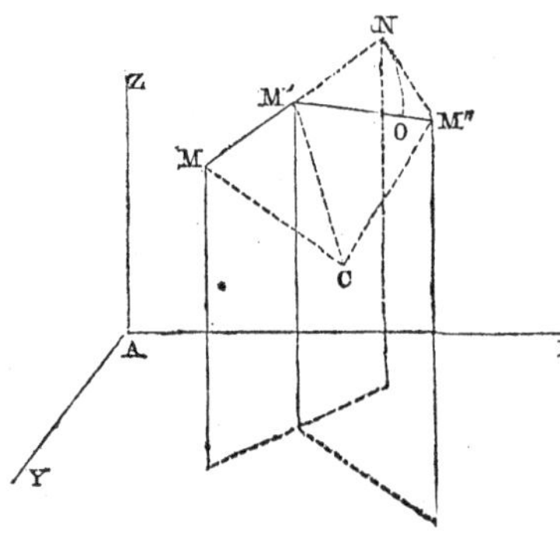

$$x + dx, \qquad y + dy, \qquad z + dz \ldots\ldots\ldots(1).$$

Differentiating these, we have

$$dx + d^2x, \qquad dy + d^2y, \qquad dz + d^2z;$$

and adding these to expressions (1) respectively, we have

$$x + 2dx + d^2x, \qquad y + 2dy + d^2y, \qquad z + 2dz + d^2z,$$

for the co-ordinates of M″. Produce the element MM′ to N, making M′N = MM′; then the co-ordinates of N will be

$$x + 2dx, \qquad y + 2dy, \qquad z + 2dz.$$

These co-ordinates of M″ and N being substituted in the general formula

$$D = \sqrt{(x - x')^2 + (y - y')^2 + (z - z')^2},$$

will give

$$M''N = \sqrt{(d^2x)^2 + (d^2y)^2 + (d^2z)^2}.$$

With M′ as a centre, describe the arc NO; then, from the elementary triangle NOM″, we have

$$NO = \sqrt{\overline{M''N}^2 - \overline{M''O}^2};$$

or since, the arc being denoted by s, MM′ = ds and M″O = d^2s,

$$NO = \sqrt{(d^2x)^2 + (d^2y)^2 + (d^2z)^2 - (d^2s)^2}.$$

The elementary triangle NM′O being similar to MM′C, we have the proportion,

$$M'N : MC :: NO : MM',$$

or

$$ds : R :: NO : ds;$$

whence

$$R = \frac{ds^2}{NO} = \frac{ds^2}{\sqrt{(d^2x)^2 + (d^2y)^2 + (d^2z)^2 - (d^2s)^2}} \dots (2),$$

the expression sought, in which either variable may be regarded as the independent one.

If the curve become of single curvature, its plane may be taken as the plane XY; z will be equal to 0, and the above expression reduce to

$$R = \frac{ds^2}{\sqrt{(d^2x)^2 + (d^2y)^2 - (d^2s)^2}} \dots\dots\dots (3),$$

the most general expression for the radius of an osculatory circle referred to rectangular co-ordinate axes in its own plane.

If in this, s be regarded as the independent variable, $d^2s = 0$, and the expression becomes

$$R = \frac{ds^2}{\sqrt{(d^2x)^2 + (d^2y)^2}}.$$

If y be the independent variable, $d^2y = 0$, and expression (3) becomes

$$R = \frac{ds^2}{\sqrt{(d^2x)^2 - (d^2s)^2}} \dots\dots\dots\dots\dots (4).$$

But

$$ds^2 = dx^2 + dy^2;$$

differentiating this, dy being constant, we have

$$2ds\,d^2s = 2dx\,d^2x, \qquad \text{or} \qquad d^2s = \frac{dx\,d^2x}{ds}.$$

Substituting this in (4), and reducing, we obtain

$$R = \frac{(ds)^3}{dy\,d^2x}.$$

If x be the independent variable, a similar process will give

$$R = \frac{(ds)^3}{dx d^2 y},$$

the same expression as in Art. (99), since in this case s represents the arc.

103. Since the curve and osculatory circle at a given point have a tangent in common, they must also have the same normal; but the normal to the circle passes through its centre; the normal to the curve must then pass through this centre; or *the radius of the osculatory circle, drawn to the point of osculation, is normal to the curve.*

104. Let BB′ be any curve, and CC′ an arc of the osculatrix of the first order, at M. Since in this case $A' = 0$, the expression for M′M″, Art. (95), becomes

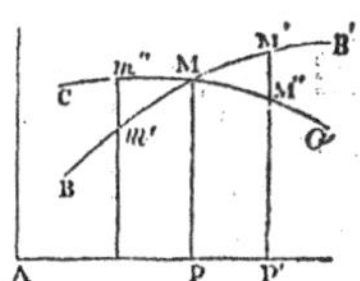

$$M'M'' = A'' \frac{h^2}{1.2} + A''' \frac{h^3}{1.2.3} + \&c. \ldots\ldots (1);$$

in which, h being infinitely small, the sign of M′M″ will be the same as that of A″, whether h be positive or negative; that is, M′M″ and $m'm''$ will have the same sign; hence, if M″ is below the curve BB′, m'' will also be below, and the converse; and the osculatrix *cannot intersect the curve at* M.

If the contact be of the second order, we have also $A'' = 0$, and

$$M'M'' = A''' \frac{h^3}{1.2.3} + A'''' \frac{h^4}{1.2.3.4} + \&c.,$$

which will have the same sign as A‴ when h is positive, and a contrary sign when h is negative; that is, $M'M''$ and $m'm''$ have contrary signs; hence, if M'' is below the curve BB', m'' will be above it, and the converse; *and the osculatrix must cut the curve at* M.

It may be shown, in the same way, that *any osculatrix of an even order intersects the curve at the point of osculation; while one of an uneven order does not.* As, when the order of contact is even, the first term of (1) will contain h with an odd exponent, and will therefore change its sign when h becomes $-h$. This will not be the case when the exponent of h in the first term of (1) is an even number.

The osculatory circle, however, does not intersect at those points about which the curve is symmetrical with its normal. For, ordinates being drawn from the points of both, perpendicular to the common normal, if the ordinate of the curve on one side is greater than the corresponding ordinate of the circle, it will be so on the other side; as may be seen in the figure, in which, if $pn > po$, then $pn' > po'$; or if $pn < po$, then $pn' < po'$; hence, in this case, in the vicinity of the point M, the circle lies entirely within or entirely without the curve. In these cases it will be found that the order of contact of the circle is odd, and higher than the second; for, unless $A''' = 0$, the circle must intersect, as shown by the preceding demonstration.

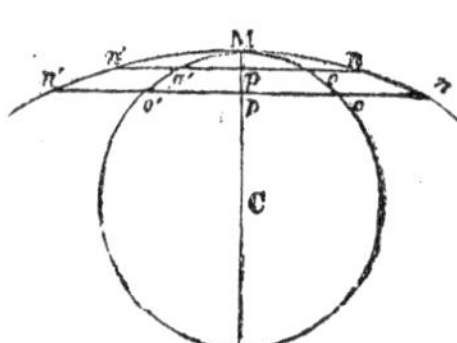

Since the osculatory circle has a more intimate contact with a curve at a given point than any other circle, it will necessarily separate those circles which are tangent without the curve from those which are tangent within.

105. *The curvature of a curve at a given point is its tendency to depart from its tangent at that point;* or *is the angular space included between the curve and its tangent.* Thus, of the two curves

AC and AB, having the common tangent AD, the former has a greater tendency to depart from the tangent, and has the greatest curvature, since the angular space DAC > DAB.

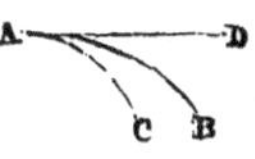

The curvature of the circumference of a circle is evidently the same at all of its points; but of two different circumferences, that one curves the most which has the least radius; as in the figure, the tendency of abd to depart from the tangent is greater than that of $ab'd'$, and this tendency plainly increases as the radius decreases, and the reverse; that is, *the curvature in two different circles varies inversely as their radii.*

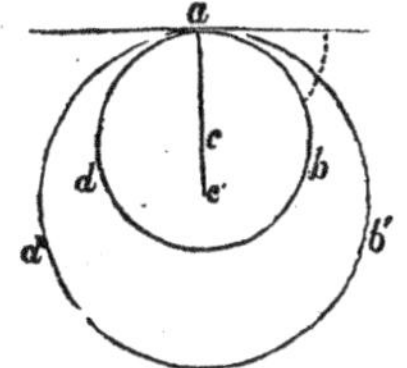

This being the case, the expression $\frac{1}{R}$ may be taken as the measure of the curvature of a circle whose radius is R.

Since the contact of the osculatory circle with a curve is so intimate, its curvature may be taken for the curvature of the curve at the point of osculation; and *the two in the immediate vicinity of this point may be regarded as one and the same curve;* hence, to compare the curvatures at different points of a curve, we have only to compare the curvatures of the osculatory circles drawn at these points. Thus, in the curve MM′,

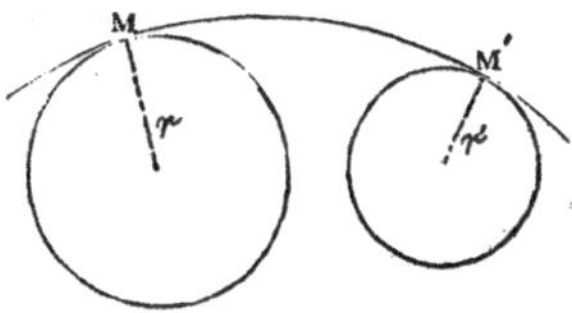

$$\textit{curvature at } \mathrm{M} : \textit{curvature at } \mathrm{M}' :: \frac{1}{r} : \frac{1}{r'}.$$

106. The radius of the osculatory circle at a given point of a curve is called *the radius of curvature*, at that point; and the centre of the circle is *the centre of curvature*. A general expression for this radius is given in article (99), and it may be found for any particular curve by differentiating the equation of the curve, and substituting the derived expressions for dy and d^2y in the formula,

$$R = \pm \frac{(dx^2 + dy^2)^{\frac{3}{2}}}{dxd^2y}.$$

If the value at any particular point of the curve be required, for x and y in the expression just deduced, substitute the co-ordinates of the particular point.

As only the absolute length of the radius of curvature is required in determining the curvature of curves, we may use either the plus or minus sign of the formula. It is best, in general, to use that which, taken with the sign resulting from the expression, will make R essentially positive.

Let it now be required to find the general expression for the radius of curvature of Conic Sections.

Their equation is

$$y^2 = 2px + r^2x^2; \quad \text{whence} \quad dy = \frac{(p + r^2x)\,dx}{y},$$

$$dx^2 + dy^2 = \frac{[y^2 + (p + r^2x)^2]\,dx^2}{y^2},$$

$$d^2y = \frac{r^2y\,dx^2 - (p + r^2x)\,dx\,dy}{y^2} = \frac{[r^2y^2 - (p + r^2x)^2]\,dx^2}{y^3}.$$

These expressions substituted in the formula give, after reduction,

$$R = \frac{[2px + r^2x^2 + (p + r^2x)^2]^{\frac{3}{2}}}{p^2} \quad \ldots\ldots\ldots(1);$$

the numerator of which is the cube of the normal, Art. (85). Hence, the radius of curvature at any point of a conic section, is equal to *the cube of the normal divided by the square of half the parameter*, and the radii at different points are to each other as the cubes of the corresponding normals.

If in (1) we make $x = 0$, we have, at the principal vertex,

$$R = p = \textit{one-half the parameter},$$

which for the ellipse and hyperbola is $\frac{b^2}{a}$.

The radius of curvature at the vertex of the conjugate axis of the ellipse, is obtained by substituting in (1),

$$p = \frac{b^2}{a}, \qquad r^2 = -\frac{b^2}{a^2}, \qquad \text{and} \qquad x = a.$$

The result is

$$R = \frac{a^2}{b} = \textit{one-half the parameter of the conjugate axis.}$$

It may be readily shown that p is *the least value* of R; therefore the curvature at the principal vertex of a conic section, is greater than at any other point. Likewise, $\frac{a^2}{b}$ is *the greatest value* of R in the ellipse; hence, the curvature of the ellipse *is least* at the vertex of the conjugate axis. The curvature of the other two curves diminishes as we recede from the vertex.

For the parabola $r^2 = 0$, we then have

$$R = \frac{(2px + p^2)^{\frac{3}{2}}}{p^2}.$$

Evolutes.

107. If, at the different points of a given curve, osculatory circles be drawn, and a second curve traced through their centres, the latter is called *the evolute* of the former, which is *the involute.* Thus, CC″ is *the evolute* of the involute MM″. Points of the evolute may always be constructed by drawing normals at the different points of the involute, and on each of these normals laying off the corresponding value of R, deduced as in article (106).

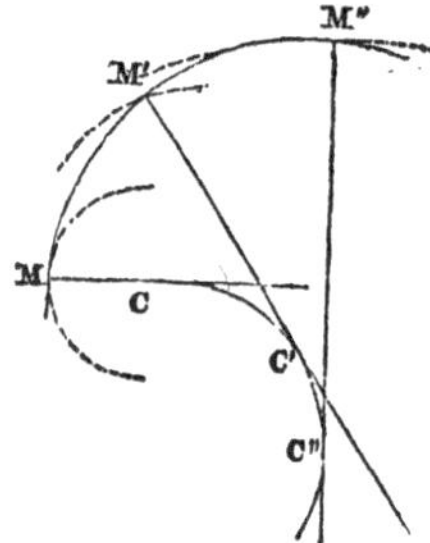

108. If α and β, the co-ordinates of the centre of the osculatory circle, be regarded as variables, they will determine all the points of the evolute; but α, β, and R, are functions of x and y, the co-ordinates of the points of osculation; and the relation between these five variables is expressed by the three equations of Art. (99), which may be written thus,

$$(x-\alpha)^2 + (y-\beta)^2 = R^2 \ldots .(1),$$

$$(x-\alpha)\,dx + (y-\beta)\,dy = 0 \ldots ..(2),$$

$$(y-\beta)\,d^2y + dy^2 + dx^2 = 0 \ldots ..(3).$$

If we differentiate (1) and (2), regarding all the quantities, except dx, as variables, we obtain

$$(x-\alpha)\,dx + (y-\beta)dy - (x-\alpha)\,d\alpha - (y-\beta)d\beta = RdR,$$

$$dx^2 + dy^2 + (y-\beta)\,d^2y - dxd\alpha - dyd\beta = 0,$$

and these, by means of equations (2) and (3), are reduced to

$$-(x-\alpha)\,d\alpha - (y-\beta)\,d\beta = \mathrm{R}d\mathrm{R} \ldots\ldots (4),$$

$$-dxd\alpha - dyd\beta = 0 \ldots\ldots\ldots (5).$$

Equation (5) gives

$$-\frac{dx}{dy} = \frac{d\beta}{d\alpha} \ldots\ldots\ldots\ldots\ldots\ldots (6).$$

$-\frac{dx}{dy}$ is the tangent of the angle which a normal to the involute at the point (x, y) makes with the axis of X, Art. (84), and $\frac{d\beta}{d\alpha}$ is the tangent of the angle which a tangent to the evolute at the point (α, β) makes with the same axis; hence, these angles are equal. But the normal at the point (x, y) passes through the point (α, β), Art. (103); therefore the normal and tangent form one and the same line; that is, *the radius of curvature is normal to the involute, and tangent to the evolute.*

The evolute may therefore be constructed by drawing a curve tangent to the normals at the different points of the involute.

From what precedes, it is plain that the evolute may be regarded as formed by the intersections of the consecutive normals to the involute, and that the point of intersection of any two consecutive normals may be taken as the centre of the osculatory circle, which passes through the two consecutive points of the involute at which the normals are drawn.

109. Equation (6) of the preceding article, combined with (2), gives

$$x - \alpha = \frac{d\alpha}{d\beta}(y-\beta).$$

Substituting this value in (1), we have, after reduction,

$$(y - \beta)^2 \frac{(d\alpha^2 + d\beta^2)}{d\beta^2} = R^2 \ldots\ldots (7).$$

Substituting the same value in (4), reducing and squaring both members, we obtain

$$(y - \beta)^2 \frac{(d\alpha^2 + d\beta^2)^2}{d\beta^2} = R^2 dR^2.$$

Dividing this by (7), member by member, and taking the root,

$$\sqrt{d\alpha^2 + d\beta^2} = dR.$$

But if z represent the arc of the evolute, we have

$$dz = \sqrt{d\alpha^2 + d\beta^2} \ldots\ldots\ldots \text{Art. (90)};$$

hence

$$dR = dz, \quad \text{and} \quad R = z + c \ldots\ldots \text{Art. (16)}.$$

110. If any two radii of curvature be drawn, as one at M′ and the other at M″; the first being denoted by R, the second by R′, and the corresponding arcs CC′ and CC″ by z and z', we have

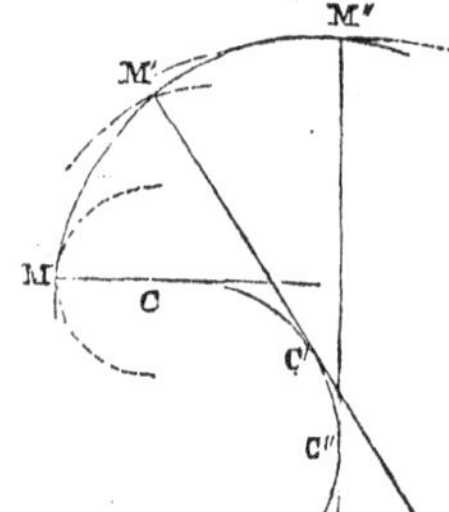

$$R = z + c, \qquad R' = z' + c;$$

whence

$$R' - R = z' - z;$$

that is, *the difference between any two radii of curvature is equal to the arc of the evolute intercepted between them.*

If in the equation $R = z + c$, we make $z = 0$, and denote by r, the corresponding value of R, we shall have

$$r = 0 + c = c;$$

that is, the constant c is always equal to the radius of curvature which passes through the point of the evolute, from which its arc is estimated.

If we estimate the evolute of the ellipse from the point C, we have

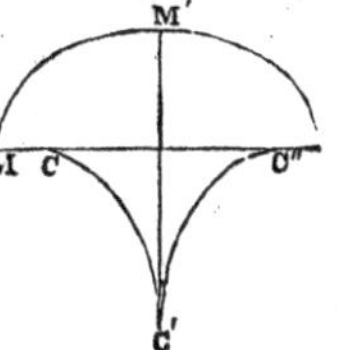

$$c = \text{MC} = \frac{b^2}{a} \ldots\ldots\ldots \text{Art. (106)};$$

hence

$$R = z + \frac{b^2}{a}.$$

Also, since $$\text{M}'\text{C}' = \frac{a^2}{b},$$

$$\text{M}'\text{C}' - \text{MC} = \frac{a^2}{b} - \frac{b^2}{a} = \text{CC}'.$$

If the evolute and one point of the involute be given, and a thread be wound upon the evolute and drawn tight, passing through the given point M, when the thread is unwound or *evolved*, the point of a pencil first placed at M will describe the involute; for by the nature of the operation, CC′ is always equal to M′C′ − MC.

111. The equation of the evolute of any curve may be found thus: Differentiate the equation of the involute twice; deduce the expressions for dy and d^2y, and substitute in the equations (2) and (3), Art. (99),

$$x - \alpha = -\frac{dy}{dx}(y - \beta)\ldots\ldots..(1),$$

$$y - \beta = -\frac{dx^2 + dy^2}{d^2y}\ldots\ldots..(2),$$

combine the results, which will contain the four variables, α, β x, and y, with the equation of the involute, and eliminate x and y; the final equation will contain only α, β, and constants, and will therefore be the required equation.

As an example, let it be required to find the equation of the evolute of the common parabola.

The equation of the involute is

$$y^2 = 2px, \qquad \text{whence} \qquad \frac{dy}{dx} = \frac{p}{y},$$

$$dy^2 = \frac{p^2dx^2}{y^2}, \qquad d^2y = -\frac{p^2dx^2}{y^3}.$$

Substituting these values in (1) and (2), and reducing, we have

$$x - \alpha = -\frac{y^2}{p} - p\ldots\ldots..(3),$$

$$y - \beta = \frac{y^3}{p^2} + y, \qquad \text{whence} \qquad -\beta = \frac{y^3}{p^2}\ldots.(4);$$

and putting for y, in (3) and (4), its value $\sqrt{2px} = (2p)^{\frac{1}{2}}x^{\frac{1}{2}}$. we have

$$x - \alpha = -2x - p, \qquad -\beta = \frac{2^{\frac{3}{2}}x^{\frac{3}{2}}}{p^{\frac{1}{2}}}.$$

The value of $x = \frac{1}{3}(\alpha - p)$ taken from the first equation, and substituted in the last, gives

$$\beta^2 = \frac{8}{27p}(\alpha - p)^3,$$

which is the required equation.

If we make $\beta = 0$, we have $\alpha = p$, and laying off $AC = p$, C will be the point at which the evolute meets the axis of X. If we transfer the origin of co-ordinates to this point, we have

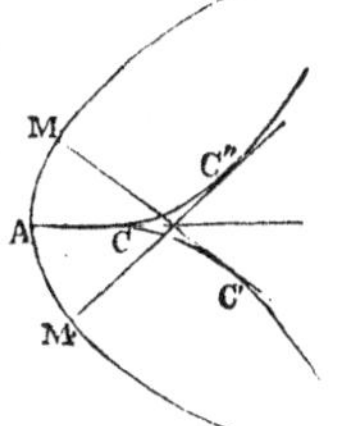

$$\alpha = p + \alpha', \qquad \beta = \beta';$$

hence

$$\beta'^2 = \frac{8}{27p}\alpha'^3.$$

Since every value of α' gives two values of β', equal with contrary signs, the curve is symmetrical with the axis of X. If α' be negative, β' is imaginary, and the curve does not extend to the left of C. The branch CC′ belongs to AM, and CC″ to AM′.

Envelopes.

112. Let $$u = f(x, y, a) = 0 \ldots\ldots\ldots (1),$$

represent a curve given in kind only, a being the only *arbitrary constant* in the equation. If a be regarded as variable, and be changed so as not to change the form of the equation, we may obtain an infinite number of curves of the same species as that represented by (1). If a be increased by da, we shall evidently obtain the curve of the species which is consecutive with the first. By increasing a again by da, we shall obtain the next consecutive curve, and so on. In general, these consecutive curves will intersect each other two and two, and by their intersections form a new curve, which is called *the envelope* of the species represented by equation (1).

113. To explain the mode of obtaining the equation of this envelope, we substitute, in equation (1) of the preceding article, $a + da$ for a, and obtain

$$u' = f(x, y, a + da) = 0;$$

or by Taylor's theorem, Art. (35),

$$u + \frac{du}{da}da + \frac{d^2u}{da^2}\frac{da^2}{1.2} + \&c. = 0 \ldots\ldots\ldots(1).$$

Since $$u = f(x, y, a) = 0,$$

and since, da being infinitely small, all terms after $\frac{du}{da}da$ may be rejected, equation (1) becomes

$$\frac{du}{da}da = 0, \qquad \text{or} \qquad \frac{du}{da} = 0 \ldots\ldots\ldots(2),$$

which is the equation of the first consecutive curve.

If this be combined with equation (1) of the preceding article, the values of x and y, in the result, will be the co-ordinates of the points of intersection of these two curves; and if they be combined in such a way as to eliminate a; x and y will be the co-ordinates of the points of intersection of any two consecutive curves of the same species, or the general co-ordinates of the curve formed by these intersections. To obtain, then, the equation of the envelope of a curve given in kind, we combine its equation with its differential equation, taken with respect to the arbitrary constant, and eliminate the constant; the result will be the required equation.

114. This elimination may be effected by deducing the expression for a, in terms of x and y, from the equation

$$\frac{du}{da} = 0 \ldots\ldots\ldots\ldots (1),$$

and substituting it in equation (1) of article (112). This expression may be represented by $a = \varphi(x, y)$.

If, then, in equation (1) of Art. (112), a be regarded as equal to $\varphi(x, y)$, that equation will represent the envelope. If, under this supposition, the equation be differentiated, we have, for its differential equation,

$$\frac{du}{dx}dx + \frac{du}{dy}dy + \frac{du}{da}da = 0,$$

which, since $\frac{du}{da} = 0$, reduces to

$$\frac{du}{dx}dx + \frac{du}{dy}dy = 0;$$

an equation identical with that obtained by differentiating equation (1), Art. (112), when a is constant. The expressions for $\frac{dy}{dx}$, deduced from these two equations, will then be the same; hence, at the point of intersection of two consecutive curves, the tangent to the envelope will be the same as the tangent to the first curve; or, the envelope is tangent to the different curves of the species, hence its name.

115. We may illustrate by the following examples:

1. Deduce the equation of the envelope formed by the intersections of the consecutive right lines given by the equation

$$u = y - ax - \frac{b}{a} = 0.\dots\dots\dots(1),$$

when a varies. Differentiating with respect to a, we have

$$\frac{du}{da} = -x + \frac{b}{a^2} = 0,$$

whence

$$a = \pm\sqrt{\frac{b}{x}}.$$

Substituting this value in (1), reducing, and squaring both members, we have

$$y^2 = 4bx,$$

the equation of a parabola.

2. Deduce the equation of the envelope of the parabolas, given by the equation

$$(1 + a^2)x^2 - 2apx + 2py = 0,$$

when a varies.

116. If the equation of the curve have two constants, we may limit the species of curves represented by it, by requiring an equation of condition to exist between these constants, such as to make one dependent upon the other. In this case, the expression for one, in terms of the other, may be obtained from the equation of condition and be substituted in the equation of the curve, and then the equation of the envelope of the species be deduced as in the preceding article. Or otherwise, the given equation may be differentiated, regarding one of the constants as a function of the other; the equation of condition may also be differentiated under the same supposition, and then, by the combination of the four equations, the differential coefficient and constants be eliminated, thus giving the equation of the envelope.

In the same way, the equation of the envelope, when there is any number of constants with a proper number of equations of condition, may be determined.

Examples.

1. Find the equation of the envelope of the different positions of a right line of given length, which is moved with its extremities in two rectangular axes.

Let l be the length of the line, a and b the distances cut off from the axes of X and Y respectively. The equation of the line may be put under the form

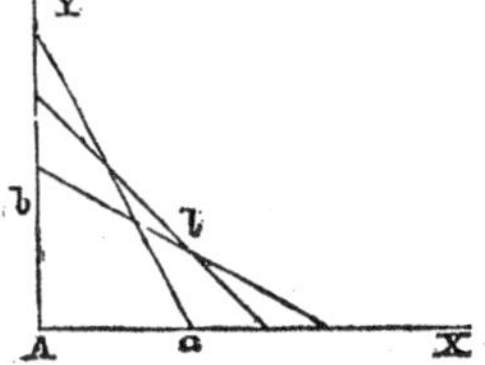

$$\frac{x}{a} + \frac{y}{b} = 1 \ldots\ldots\ldots (1).$$

From the condition of the problem, we also have

$$a^2 + b^2 = l^2 \ldots\ldots\ldots\ldots (2).$$

Differentiating, regarding a as a function of b, we have

$$\frac{x}{a^2}da + \frac{ydb}{b^2} = 0,$$

$$ada + bdb = 0.$$

Deducing from these the expressions for $\frac{da}{db}$, and equating,

$$\frac{a^2y}{b^2x} = \frac{b}{a}, \qquad \text{whence} \qquad \frac{1}{a} = \frac{a^2y}{b^3x}.$$

Substituting this in (1), we have

$$\frac{a^2y}{b^3} + \frac{y}{b} = 1, \qquad \text{or} \qquad (a^2 + b^2)y = b^3;$$

and since $a^2 + b^2 = l^2$,

$$b^3 = yl^2, \qquad b = \sqrt[3]{yl^2}.$$

In the same way, we find

$$a = \sqrt[3]{xl^2}.$$

Substituting these expressions in (2), and reducing, we have, for the equation of the envelope,

$$x^{\frac{2}{3}} + y^{\frac{2}{3}} = l^{\frac{2}{3}}.$$

2. Find the equation of the envelope of a series of ellipses having the same centre, same co-ordinate axes, and same area.

Let the equation of the ellipse be put under the form,

$$\frac{y^2}{b^2} + \frac{x^2}{a^2} = 1 \ldots\ldots\ldots (1).$$

Since the areas are the same, we must have

$$ab = c^2 \ldots\ldots\ldots\ldots (2),$$

c^2 being constant. By differentiating, &c., as in the preceding article, we find for the envelope,

$$xy = \frac{c^2}{2}$$

the equation of an equilateral hyperbola, referred to its asymptotes.

Application of the Differential Calculus to the Construction and Discussion of Curves. Singular Points.

117. The most general division of curves is into the classes, *Algebraic* and *Transcendental.*

When the relation between the ordinate and abscissa of a curve can be expressed entirely in algebraic terms, Art. (5), it belongs to the first class; and when such relation cannot be expressed without the aid of transcendental quantities, it belongs to the second class.

We have seen, in Analytical Geometry, the mode of constructing and discussing curves when their equations are given. By the aid of the Differential Calculus, this discussion may not only be simplified but much extended, and the nature, form, and properties of the curve be thus more fully ascertained.

On many curves points are found, at which there exists some remarkable property not enjoyed by the other points of the curve. These are called *singular points.* They are entirely independent of the system or position of the co-ordinate axes, and are easily discovered and characterized by the Calculus.

A detailed discussion of the general equation

$$y = b + c(x - a)^m \dots\dots\dots\dots (1),$$

in which m is any positive number, will illustrate these principles.

First: *Let m be entire and even.*

Since every value of x, positive or negative, gives one real value for y, the curve is continuous, and extends indefinitely in the direction of the axis of X.

By the differentiation, &c., of (1), we have

$$\frac{dy}{dx} = mc(x-a)^{m-1} \ldots (2), \quad \frac{d^2y}{dx^2} = m(m-1)c(x-a)^{m-2} \ldots (3),$$

. . . .

$$\frac{d^m y}{dx^m} = m(m-1) \ldots\ldots 2 \,.\, 1 \,.\, c.$$

Placing $\frac{dy}{dx} = 0$, we obtain $x = a$.

This value of x, when substituted in (1), (2), (3), &c., gives $y = b$, and reduces the successive differential coefficients to 0, as far as the mth, which, *if c be positive*, becomes a positive constant, and is of an even order; hence, $y = b$ is a minimum ordinate, Art. (70).

Since for $x = a$, we have $\frac{dy}{dx} = 0$, the tangent line at the extremity of this minimum ordinate is parallel to the axis of X; and since (m and $m - 2$ being even) for all values of x except $x = a$, y and $\frac{d^2y}{dx^2}$ are positive, the curve at all of its points is convex towards the axis of X, Art. (86).

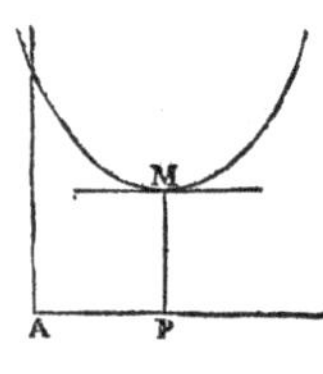

If c be negative; the mth differential coefficient will be negative; and $x = a$ and $y = b$ will be the co-ordinates of a point at which the ordinate is a maximum. In this case, the second differential coefficient for all values of x, except $x = a$, is negative, and the curve, for all positive values of y, concave, and for all negative values of y, convex, towards the axis of X.

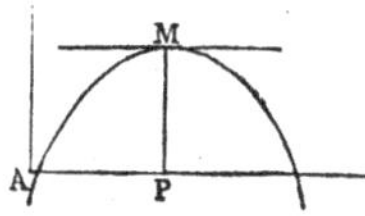

118. Second: *Let m be entire and odd.*

In this case, each value of x gives one real value for y; and each value of y, a real value for x; hence, the curve is unlimited in either direction.

When $x = a$, the first differential coefficient, as before, is equal to 0; as also the second, third, &c. The mth, however, *if c be positive*, is a positive constant, and of an odd order; there is then in this case, neither a maximum nor a minimum, Art. (70).

By examining the second differential coefficient, we see (since $m - 2$ is odd), that for every value of $x < a$, it is negative; that for $x = a$, it is 0; and when $x > a$, it is positive: hence, for all values of $x < a$, which give y positive, the curve is concave towards the axis of X; and for all values of $x > a$, it is convex, as in the figure. Therefore, at the point whose co-ordinates are $x = a$ and $y = b$, as x increases, the curve changes from being concave, and becomes convex, towards the axis of X.

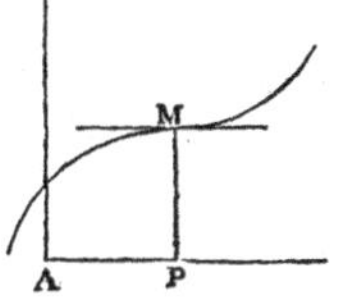

If c be negative; the reverse will be the case, and as in the second figure, at the point M, whose co-ordinates are $x = a$ and $y = b$, there is a change from convexity to concavity towards the axis of X. Such points are *singular*, and are called *points of inflexion.* In both cases the tangent line at the point of inflexion is parallel to the axis of X, and also cuts the curve.

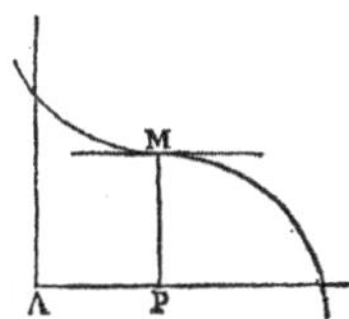

119. Third: *Let m be a fraction, the numerator and denominator of which are odd, as $\frac{3}{5}$.*

Then
$$y = b + c(x - a)^{\frac{3}{5}},$$

$$\frac{dy}{dx} = \frac{3c}{5(x-a)^{\frac{2}{5}}}, \quad ; \quad \frac{d^2y}{dx^2} = -\frac{2}{5}\frac{3c}{5(x-a)^{\frac{7}{5}}} \ldots\ldots\text{\&c.};$$

$x = a$ gives

$$y = b, \qquad \frac{dy}{dx} = \infty, \qquad \frac{d^2y}{dx^2} = \infty, \text{ \&c.}$$

If c be positive; $\frac{d^2y}{dx^2}$, for all values of $x < a$, will be positive, and for all values of $x > a$, negative; hence, for all values of x less than a which give y positive, the curve will be convex, and for all values of x greater than a it will be concave towards the axis of X, as in the figure.

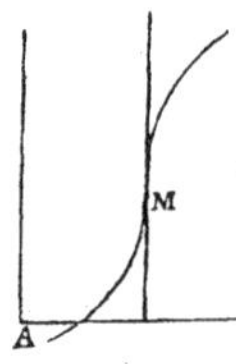

If c be negative; the reverse is the case, as in the second figure. The point M, whose co-ordinates are $x = a$ and $y = b$, is in both cases a point of inflexion at which the tangent line is perpendicular to the axis of X Whence we may say: *A point of inflexion is one at which, as the abscissa increases, a curve changes from being concave towards any right line, not passing through the point, and becomes convex, or the reverse.*

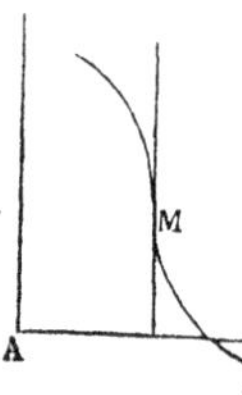

If the right line be taken as the axis of abscissas, this point will always be characterized by *a change of sign in the second differential coefficient of the ordinate.* For, since the curve on one side of the point is concave, and on the other convex, the second differential coefficient in one case has a different sign from that of the ordinate, and in the other the same; hence, at the point the sign must have changed. In order that this may be the case, the second differential coefficient must be equal to zero, or infinity.

The roots of the two equations,

$$\frac{d^2y}{dx^2} = 0, \qquad \text{and} \qquad \frac{d^2y}{dx^2} = \infty,$$

will then give all the values of the variable which can belong to points of inflexion.

It sometimes happens that a point of inflexion lies on the axis of X, as in the second case above discussed when $b = 0$. In this case $x = a$ gives

$$y = 0, \qquad \text{and} \qquad \frac{dy}{dx} = 0,$$

and the corresponding point M is a point of inflexion, at which both the second differential coefficient and ordinate change their signs.

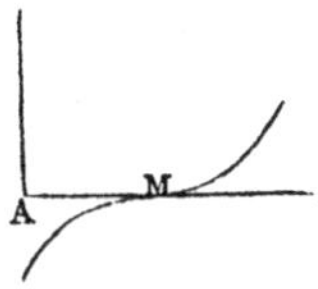

It is evident, from the preceding discussion, that if any right line be drawn through a point of inflexion, the curve on both sides of the point will either be convex towards the line, or concave.

120. Fourth: *Let m be a fraction with an even numerator, as* $\frac{2}{3}$.

Then
$$y = b + c(x - a)^{\frac{2}{3}},$$

$$\frac{dy}{dx} = \frac{2c}{3(x-a)^{\frac{1}{3}}}, \qquad \frac{d^2y}{dx^2} = -\frac{1}{3}\,\frac{2c}{3(x-a)^{\frac{4}{3}}};$$

$x = a$ gives

$$y = b, \qquad \frac{dy}{dx} = \infty, \qquad \frac{d^2y}{dx^2} = \infty.$$

If c be positive; for $x < a$, $\frac{dy}{dx}$ will be negative, and for $x > a$, it will be positive; hence at the point whose co-ordinates are $x = a$ and $y = b$, $\frac{dy}{dx}$ must change its sign from minus to plus, which change indicates a minimum ordinate, Art. (69).

If c be negative; the reverse will be the case, there will be a change of sign from plus to minus, and the ordinate will be a maximum.

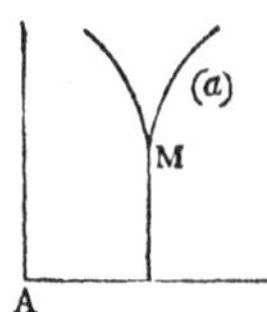

In the first case, the second differential coefficient for all values of x is negative, and the ordinate positive; the curve is therefore concave towards the axis of X, as represented in fig. (a).

In the second case, $\dfrac{d^2y}{dx^2}$ is always positive. For all positive values of y the curve will then be convex, and for all negative values of y concave, as in fig. (b). The tangent at the point M is in both cases perpendicular to the axis of X.

The point M is singular, and is called *a cusp*. It is a point at which the curve, *when interrupted in its course in one direction, turns immediately into a contrary one.*

121. Fifth: *Let m be a fraction with an even denominator, as* $\frac{3}{2}$. Since the denominator of the fraction indicates that the square root is to be taken, the double sign $\pm$ must be placed before $(x-a)^{\frac{3}{2}}$, and we then have

$$y = b \pm c(x-a)^{\frac{3}{2}},$$

$$\frac{dy}{dx} = \pm \frac{3}{2}c(x-a)^{\frac{1}{2}}, \qquad \frac{d^2y}{dx^2} = \pm \frac{3c}{4(x-a)^{\frac{1}{2}}}.$$

Every value of $x < a$ gives y imaginary; $x = a$ gives $y = b$, and $x > a$ gives two values, one greater and the other less than b. There is then no point on the left of that one whose co-ordinates are $x = a$ and $y = b$; but on the right of this point the curve must extend indefinitely, and consist of two branches.

$$x = a \quad \text{gives} \quad \frac{dy}{dx} = 0;$$

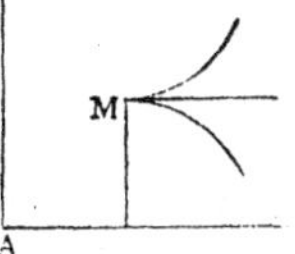

the tangent at M is then parallel to the axis of X.

Each value of $x > a$ gives two values for $\frac{d^2y}{dx^2}$, the one positive corresponding to the greater value of y, and the other negative; hence, the upper branch is convex, and the lower, until it crosses the axis of X, concave, as in the figure, and the point M is a cusp.

122. Let us now take the equation

$$(y - x^2)^2 = x^5,$$

from which we deduce

$$y = x^2 \pm x^{\frac{5}{2}},$$

$$\frac{dy}{dx} = 2x \pm \frac{5}{2}x^{\frac{3}{2}}, \qquad \frac{d^2y}{dx^2} = 2 \pm \frac{5}{2}\,\frac{3}{2}x^{\frac{1}{2}}.$$

When $x = 0$, we have $y = 0$. If x be negative, y is imaginary. For every positive value of x, there are two real values of y, both of which are positive as long as $x^2 > x^{\frac{5}{2}}$ or $x < 1$; after which, one is positive and the other negative.

When $x = 0$, $\frac{dy}{dx} = 0$; also when

$$2 \pm \frac{5}{2}x^{\frac{1}{2}} = 0, \qquad \text{or} \qquad x = \frac{16}{25};$$

hence the axis of X is tangent to the curve at the origin; and the tangent to the lower branch, at the point whose abscissa is $\frac{16}{25}$, is parallel to the axis of X.

The first value of $\frac{d^2y}{dx^2}$ belongs to the upper branch, and is always positive. The second value is also positive as long as $2 > \frac{5}{2}\frac{3}{2}x^{\frac{1}{2}}$, or $x < \frac{64}{225}$; after which it is negative.

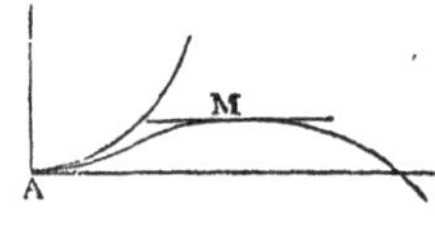

The origin is then a cusp, at which both branches lie on the same side of the common tangent, and is of the *second species*, those before discussed being of the *first species*. The point of the lower branch whose abscissa is $\frac{64}{225}$ is a point of inflexion.

123. From the equation,

$$ay^2 - x^3 + bx^2 = 0,$$

we derive

$$y = \pm\sqrt{\frac{x^2(x-b)}{a}}, \qquad \frac{dy}{dx} = \pm\frac{3x - 2b}{2\sqrt{a(x-b)}}.$$

Since $x = 0$ gives $y = 0$, the origin A is a point of the curve. All negative values of x make y imaginary, as also all positive values less than b; hence, A has no consecutive point. Such points, given by the equation of a curve, but having no consecutive points on either side, are *singular*, and are called *isolated* or *conjugate points*.

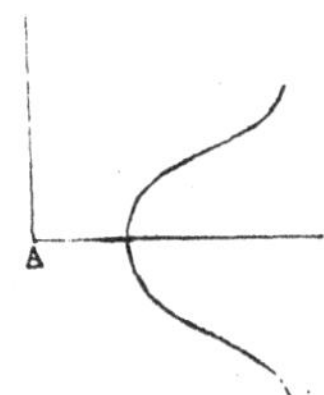

Substituting 0 for x in the expression $\frac{dy}{dx}$, it reduces to

$$\frac{\mp b}{\sqrt{-ab}},$$

an imaginary expression; and, in general, at a conjugate point, *one or more of the differential coefficients of the ordinate must be imaginary*, since y', the consecutive ordinate, when developed as in Art. (35), can only be imaginary under this supposition.

If we take the equation,

$$a^4y^2 = x^6 - a^2x^4,$$

whence

$$y = \pm \frac{x^2}{a^2}\sqrt{x^2 - a^2}, \qquad \frac{dy}{dx} = \pm \frac{3x^3 - 2a^2x}{a^2\sqrt{x^2 - a^2}}.$$

$x = 0$ and $y = 0$ will satisfy the equation, while no other value of x, numerically less than a, will give real values for y. The origin is then *a conjugate point*. In this case, for $x = 0$, $\frac{dy}{dx}$ reduces to 0. If the second differential coefficient be taken, it will, for $x = 0$, reduce to an imaginary expression.

124. Take the equation,

$$y = b \pm (x - a)\sqrt{x - c},$$

and suppose $a > c$. By differentiating, we derive

$$\frac{dy}{dx} = \pm \sqrt{x - c} \pm \frac{x - a}{2\sqrt{x - c}}.$$

For every value of $x < c$, except $x = a$, y is imaginary.

For $\quad x = c, \quad y = b, \quad$ and $\quad \frac{dy}{dx} = \infty.$

For every value of $x > c$, there are two real values of y.

For $\quad x = a, \quad y = b, \quad$ and $\quad \frac{dy}{dx} = \pm \sqrt{a - c},$

and at the corresponding point M there are two tangents, one making an angle, the tangent of which is $+\sqrt{a-c}$, and the other $-\sqrt{a-c}$. The point M is *singular*, and belongs to a class called *multiple points*, or points at which two or more branches of a curve intersect. If but two intersect, the point is a double multiple point; if three, a triple; and so on. Since there will be a separate tangent to each branch, at one of these points, it will be characterized by *two or more values of the first differential coefficient*, for the same values of the variables.

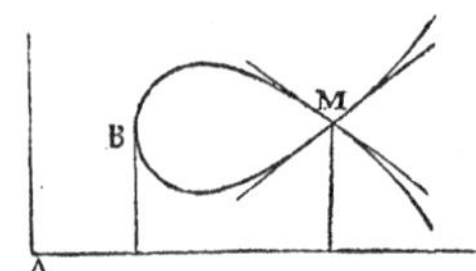

If $a < c$, $x = a$, and $y = b$ give a conjugate point.

125. We will close the discussion of algebraic curves by constructing the curve given by the equation

$$ay^2 - x^3 + (b-c)x^2 + bcx = 0;$$

whence

$$y = \pm\sqrt{\frac{x(x-b)(x+c)}{a}}, \quad \frac{dy}{dx} = \pm\frac{3x^2 - 2x(b-c) - bc}{2\sqrt{ax(x-b)(x+c)}}.$$

Each of the values, $x = 0$, $x = b$, $x = -c$, gives $y = 0$.

Every negative value of x, numerically greater than c, gives y imaginary; while every such value less than c gives two equal values of y with contrary signs. Every positive value of $x < b$ gives y imaginary; and every such value greater than b, gives two equal values of y with contrary signs. The curve is then symmetrical with reference to the axis of X.

Each of the values, $x = 0$, $x = b$, $x = -c$, reduces $\frac{dy}{dx}$ to ∞; hence, at the three corresponding points the tangent is perpendicular to the axis of X.

By solving the equation,

$$3x^2 - 2x(b - c) - bc = 0,$$

we shall find two real values for x, one positive and the other negative, and thus determine the points at which the tangent is parallel to the axis of X. The positive value will be found to be less than b, and hence will give no point of the curve. The negative value is numerically less than c, and gives two points, one above and the other below the axis of X. The curve may then be drawn as in the figure, in which $AC = -c$, and $AB = b$.

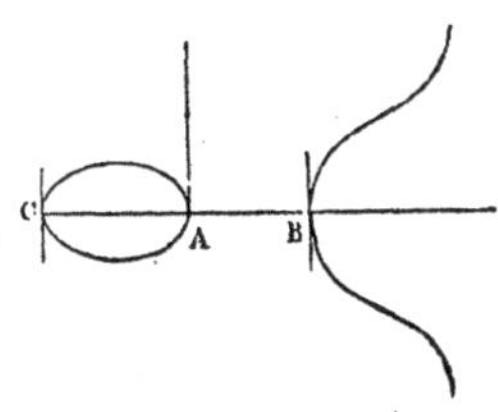

If $c = 0$, the equation becomes

$$ay^2 - x^3 + bx^2 = 0,$$

and the oval AC reduces to the conjugate point A, as in article (123).

If $b = 0$, the equation becomes

$$ay^2 - x^3 - cx^2 = 0,$$

and the curve takes the form indicated in figure (b), the origin being a double multiple point, since $\frac{dy}{dx}$ becomes equal to $\pm\sqrt{\frac{c}{a}}$.

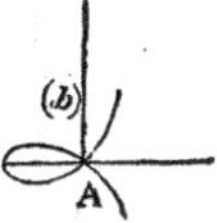

If b and c are both equal to 0, the equation becomes

$$ay^2 - x^3 = 0; \text{ whence } y = \pm\sqrt{\frac{x^3}{a}},$$

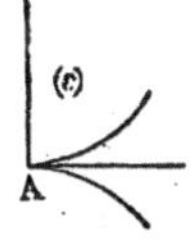

and the curve will be as in figure (c), the point A being a cusp of the first species.

126. One of the most important of the *transcendental curves* is

THE LOGARITHMIC CURVE,

so named because it may always be referred to a set of co-ordinate axes, such that one co-ordinate will be the logarithm of the other. Its equation is usually written

$$y = \log x,$$

or, if a be the base of the system of logarithms,

$$x = a^y.$$

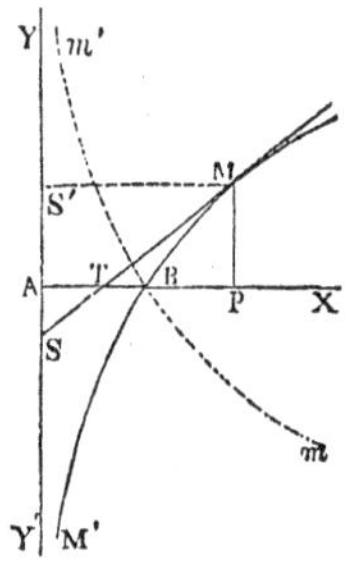

The curve is given when a is known, and may be constructed by laying off on the axis of X the different numbers, and on the corresponding perpendiculars the logarithms of these numbers. Or it may be constructed from the equation $x = a^y$, by making $y = \frac{1}{2}$, $\frac{3}{2}$, $\frac{1}{4}$, &c.; whence, the corresponding values of x are

$$x = \sqrt{a}, \quad x = a\sqrt{a}, \quad x = \sqrt[4]{a}, \text{ \&c.}$$

When $y = 0$, $x = 1$. This being the case for all systems of logarithms, shows that all logarithmic curves, when referred to the same axes, cut the axis of X, or *axis of numbers*, at a distance from the origin equal to unity.

If $a > 1$, and $x > 1$, y is positive, and increases as x increases; if $x < 1$, y is negative, and increases numerically as x decreases, until $x = 0$, when $y = -\infty$. If x be negative, there will be no corresponding value of y. The curve will then be of the form indicated by the full line in the figure.

If $a < 1$, the reverse will be the case, and the curve will be represented by the dotted line.

127. If now we differentiate the equation $y = \log x$, M being the modulus, we deduce

$$\frac{dy}{dx} = \frac{M}{x}, \qquad \frac{d^2y}{dx^2} = -\frac{M}{x^2}.$$

When $\quad x = 0, \qquad \frac{dy}{dx} = \frac{M}{0} = \infty;$

hence, the tangent at the corresponding point is the axis of Y; and since for $x = 0$, $y = -\infty$, this tangent is an asymptote.

When $\quad x = \infty, \qquad \frac{dy}{dx} = \frac{M}{\infty} = 0.$

But $x = \infty$ gives $y = \infty$; hence, there is no tangent parallel to the axis of X, at a finite distance from it.

The value for the subtangent on the axis of X is

$$PT = y\frac{dx}{dy} = \log x \frac{x}{M}.$$

If the subtangent be taken on the axis of Y, we have

$$SS' = x\frac{dy}{dx} = M;$$

that is, *the subtangent on the axis of logarithms is constant*, and equal to the modulus of the system in which the logarithms are taken.

If $M = 1, \qquad SS' = 1 = AB.$

Since, when $a > 1$, $\frac{d^2y}{dx^2}$ is negative for all values of x, the part BM is concave towards the axis of X, and BM′ convex.

When $a < 1$, M is negative, $\frac{d^2y}{dx^2}$ will be positive, the part Bm' convex, and Bm concave.

128. The curve given by the equation

$$y = x\,l x,$$

is remarkable. Each value of x gives but a single value of y. For values of $x > 1$, y is positive; for values of $x < 1$, y is negative; and for small values of x, decreases numerically and approaches the limit 0, which it reaches when $x = 0$. Negative values of x give no values for y. The origin is then a point of the curve at which it is interrupted in its course, but does not turn into a contrary one as at a cusp. Such points are called *terminating points.*

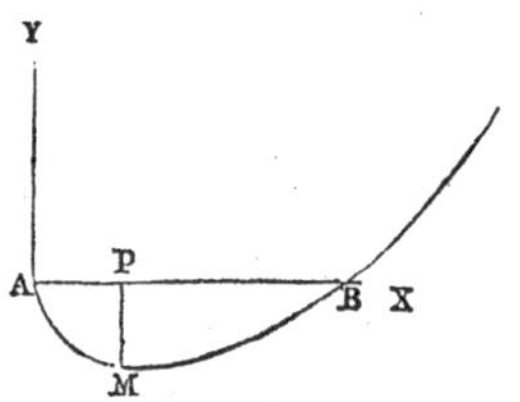

The value $x = 1$ gives $y = 0$; hence, the curve cuts the axis of X at a distance from the origin equal to 1.

Differentiating the equation, we have

$$\frac{dy}{dx} = lx + 1, \qquad \frac{d^2y}{dx^2} = \frac{1}{x}.$$

Placing $\quad lx + 1 = 0, \quad$ we have $\quad lx = -1,$

or,

$$x = e^{-1} = \frac{1}{e} = \frac{1}{2,71.....};$$

which corresponds to a minimum value of y, Art. (70).

At the point B, $\frac{dy}{dx}$ becomes equal to 1, and the tangent makes an angle of 45° with the axis of X. Between the points A and B the curve is concave, and from the point B it is convex towards the axis of X, Art. (86).

129. Let

$$y = e^{-\frac{1}{x}}.$$

Each value of x gives a single positive value of y. $x = \infty$ gives $y = 1$. As x decreases, y decreases, until $x = 0$ gives

$y = 0$. If x be negative and infinitely small, y is infinitely great, and as x increases numerically, y decreases, until $x = -\infty$ gives $y = 1$. At the origin the curve is interrupted, as in the preceding article.

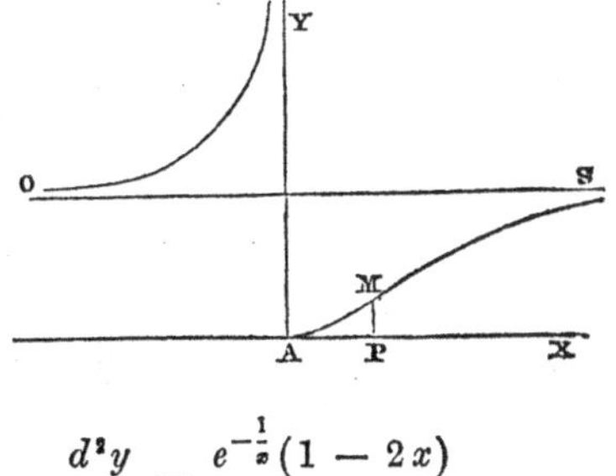

Differentiating, we have

$$\frac{dy}{dx} = \frac{e^{-\frac{1}{x}}}{x^2}, \qquad \frac{d^2y}{dx^2} = \frac{e^{-\frac{1}{x}}(1 - 2x)}{x^4}.$$

For $x = 0$, $\frac{dy}{dx}$ becomes ∞, and the axis of Y is an asymptote of the left-hand branch of the curve. For $x = \infty$ or $-\infty$, $\frac{dy}{dx}$ becomes 0, and the line OS at a distance from the axis of X equal to 1, is an asymptote to both branches. For all negative values of x, $\frac{d^2y}{dx^2}$ is positive, and the curve convex towards the axis of X. Also, for all positive values of $x < \frac{1}{2}$. For $x = \frac{1}{2}$, $\frac{d^2y}{dx^2}$ becomes 0, and is negative for all values of $x > \frac{1}{2}$. The point M, whose abscissa is $\frac{1}{2}$, is a point of inflexion, Art. (119), and beyond this point the curve is concave towards the axis of X.

130. Another singular point is given by the equation,

$$y = b + x \tan^{-1}\frac{1}{x},$$

whence

$$\frac{dy}{dx} = \tan^{-1}\frac{1}{x} - \frac{x}{x^2 + 1}.$$

Each value of x gives but a single value for y. For values of x which are numerically equal, one positive and the other negative, the corresponding values of y are equal; hence, the curve is symmetrical with respect to the axis of Y, and $x = 0$ gives $y = b$.

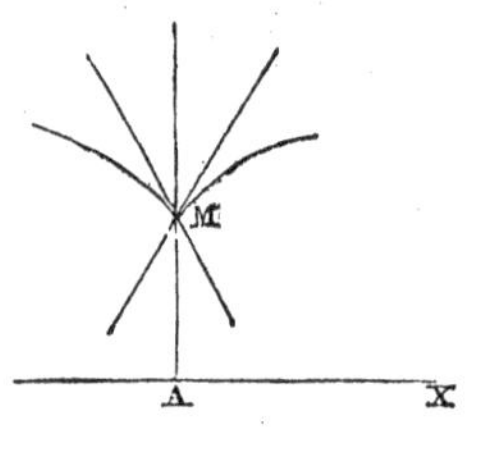

For all positive values of x, $\frac{dy}{dx}$ is positive, and as x is diminished to 0, $\frac{dy}{dx}$ increases to

$$\left(\frac{dy}{dx}\right)_{x=0} = \tan^{-1}\frac{1}{0} = \tan^{-1}\infty = \frac{\pi}{2}.$$

For negative values of x, $\frac{dy}{dx}$ is negative, and increases numerically as x is thus decreased to 0, when we have

$$\left(\frac{dy}{dx}\right)_{x=0} = \tan^{-1}\left(-\frac{1}{0}\right) = \tan^{-1}(-\infty) = -\frac{\pi}{2}.$$

We thus have two branches terminating at the point M, not tangent to each other as at a cusp. This, which is but a particular case of a multiple point, is called *a salient point.*

131. The most remarkable transcendental curve is

THE CYCLOID,

which is generated by a point in the circumference of a circle, when the circle is rolled in the same plane, along a given straight line.

Let AB be the given line, and suppose the circle to have been placed upon it, so that the generating point was at A, and then to have been rolled to the position RME.

The generating point now at M, has generated the arc AM.

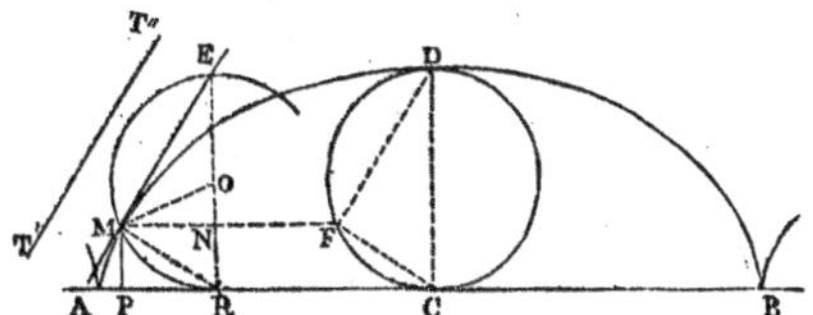

Take the origin of co-ordinates at A, and let $AP = x$, $PM = y$ and RE, the diameter of the generating circle $= 2r$; then

$$AP = AR - PR \ldots\ldots\ldots (1).$$

But since every point of the circumference from M to R, as the circle was rolled, came in contact with AR, we have

$$AR = \text{arc } MR = \text{ver-sin}^{-1} RN = \text{ver-sin}^{-1} y.$$

Also,

$$PR = MN = \sqrt{RN \times NE} = \sqrt{y(2r - y)} = \sqrt{2ry - y^2}.$$

Substituting the values of AP, AR, and PR in (1), we have

$$x = \text{ver-sin}^{-1} y - \sqrt{2ry - y^2} \ldots\ldots\ldots (2),$$

which is the equation of the Cycloid.

After the circle has been rolled over once, every point of the circumference will have been in contact with AB, and the generating point will have arrived at B; we have then

$$AB = \textit{circumference of generating circle} = 2\pi r.$$

The given line is called the base of the Cycloid, and the line CD $= 2r$ perpendicular to AB at its middle point, is the axis.

If the rolling of the circle be continued beyond the point B, an infinite number of arcs, each equal to ADB, will be generated.

Every negative value of y in equation (2) makes x imaginary; hence there is no point of the curve below the axis of X.

$$y = 2r, \qquad \text{gives} \qquad x = \text{ver-sin}^{-1} 2r = \pi r = \text{AC}.$$

Every value of $y > 2r$ makes x imaginary; hence the greatest ordinate of the curve is equal to the diameter of the generating circle. For the points of each branch between D and B, the essential sign of the radical must evidently be plus.

By differentiating (2) we have, Art. (44), after introducing the radius r,

$$dx = \frac{rdy}{\sqrt{2ry - y^2}} - \frac{rdy - ydy}{\sqrt{2ry - y^2}};$$

or reducing

$$dx = \frac{ydy}{\sqrt{2ry - y^2}} \dots\dots\dots (3),$$

which is the differential equation of the Cycloid.

132. Substituting the preceding value of dx in the formulas of article (85), and reducing, we have

$$\textit{Subtangent}, \text{ PT} = \frac{y^2}{\sqrt{2ry - y^2}}.$$

$$\textit{Tangent}, \text{ MT} = \frac{y\sqrt{2ry}}{\sqrt{2ry - y^2}}.$$

$$\textit{Subnormal}, \text{ PR} = \sqrt{2ry - y^2}.$$

Normal, $MR = \sqrt{2ry}$.

Since the subnormal $PR = \sqrt{2ry - y^2} = MN$, the diameter ER and normal MR intersect the base at the same point. Hence, to construct the normal at a given point, join it with the point at which the corresponding position of the generating circle is tangent to the base. Or, upon the greatest ordinate CD as a diameter, describe a circle, and through the given point M draw a line parallel to the base; from the point F in which it cuts the circle, draw the two chords CF and DF to the extremities of the diameter; a line through the given point parallel to CF will be the normal, and one parallel to DF the tangent.

If it be required to draw a tangent parallel to a given line, as T′T″, draw the chord DF parallel to the given line, from F draw FM parallel to the base; the point M is the point of contact, through which draw a line parallel to T′T″.

133. From equation (3), article (131), we have

$$\frac{dy}{dx} = \frac{\sqrt{2ry - y^2}}{y} = \sqrt{\frac{2r}{y} - 1}\ldots\ldots\ldots(1),$$

which becomes 0 when $y = 2r$, and ∞ when $y = 0$; hence, at the extremity of the greatest ordinate, the tangent is parallel to the base; and at the points A, B, &c., where the curve meets the base, it is perpendicular.

If we square both members of equation (1), we have

$$\frac{dy^2}{dx^2} = \frac{2r}{y} - 1.$$

Differentiating both members of this, we have

$$\frac{2dyd^2y}{dx^2} = -\frac{2rdy}{y^2}, \qquad \text{or} \qquad \frac{d^2y}{dx^2} = -\frac{r}{y^2}.$$

This second differential coefficient being negative for all values of y, the curve is concave towards the axis of X, Art. (86).

134. Substituting the values of dy and d^2y in the expression

$$R = -\frac{(dx^2 + dy^2)^{\frac{3}{2}}}{dxd^2y},$$

we obtain

$$R = \frac{\left(\frac{2\,rydx^2}{y^2}\right)^{\frac{3}{2}}}{\frac{rdx^3}{y^2}} = 2^{\frac{3}{2}}r^{\frac{1}{2}}y^{\frac{1}{2}} = 2\sqrt{2ry};$$

or since $\sqrt{2ry}$ is the expression for the normal, Art. (132), *the radius of curvature is equal to twice the normal at the point of osculation.*

If $y = 0$, $R = 0$; and if $y = 2r$, $R = 4r$;

hence, the radius of curvature at A (see figure in next article) is equal to 0; and at D is $4r$; therefore, Art. (110), the arc $AA' = 4r$.

135. To obtain the equation of the evolute, let us substitute the values of dy and d^2y in equations (1) and (2) of article (111).

After reduction, we find

$$y - \beta = 2y, \qquad x - \alpha = -2\sqrt{2ry - y^2};$$

whence

$$y = -\beta, \qquad x = \alpha - 2\sqrt{-2r\beta - \beta^2}.$$

These values, in the equation of the involute, Art. (131), give

$$\alpha = \text{ver-sin}^{-1}(-\beta) + \sqrt{-2r\beta - \beta^2}\dots\dots(1),$$

for the required equation.

If we produce DC to A′, making CA′ = DC, and then transfer the origin to A′, the new axes being A′X′ and A′D, and the new co-ordinates α' and β', we shall have for any point, as M′,

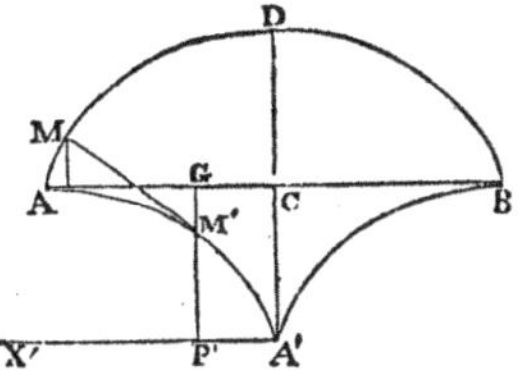

$$AG = \alpha, \qquad GM' = -\beta,$$

$$A'P' = \alpha', \qquad P'M' = \beta'.$$

Since AC $= \pi r$, and CG = A′P′,

$$\alpha = \pi r - \alpha';$$

and since GP′ $= 2r$,

$$GM' = 2r - \beta', \qquad \text{or} \qquad -\beta = 2r - \beta'.$$

Substituting these values in (1), we have

$$\pi r - \alpha' = \text{ver-sin}^{-1}(2r - \beta') + \sqrt{2r\beta' - \beta'^2},$$

whence

$$\alpha' = \pi r - \text{ver-sin}^{-1}(2r - \beta') - \sqrt{2r\beta' - \beta'^2};$$

But $\pi r - \text{ver-sin}^{-1}(2r - \beta') = \text{ver-sin}^{-1}\beta'$; hence, the last equation becomes

$$\alpha' = \text{ver-sin}^{-1}\beta' - \sqrt{2r\beta' - \beta'^2},$$

which is the equation of the evolute referred to the new axes, and is of the same form and contains the same constants as the equation of the involute; therefore the two curves are equal.

Since arc $AA' = 4r$, its equal $AD = 4r$, and $ADB = 4.2r$; that is, *equal to four times the diameter of the generating circle.*

Polar Curves. Spirals.

136. In Analytical Geometry we have seen, that we may obtain the polar equation of any curve, given in terms of rectangular co-ordinates, by substituting for these co-ordinates their values, in terms of the polar co-ordinates, taken from the formulas

$$x = a + r\cos v, \qquad y = b + r\sin v \ldots . (1).$$

Also, if we have the differential equation of the curve, or any expression containing the differentials of the variables, we at once pass to the corresponding equation or expression in terms of polar co-ordinates and their differentials, by substituting for x and y the above expressions, and for dx and dy the expressions below, obtained by differentiating equations (1).

$$dx = \cos v\,dr - r\sin v\,dv,$$

$$dy = \sin v\,dr + r\cos v\,dv.$$

137. If a right line be revolved uniformly, in the same plane, about one of its points, a second point of the line continually approaching, or receding from the fixed point, in accordance with some prescribed law, will generate a curve called *a spiral.*

The fixed point is called the *pole* or eye of the spiral. The portion of the spiral generated while the line makes one revolution, is called *a spire;* and since there is no limit to the number of revolutions, the number of spires is infinite, and any straight line drawn through the pole of the spiral will intersect it in an infinite number of points. For this reason, the relation between the ordinate and abscissa of a spiral cannot be expressed algebraically, Art. (117).

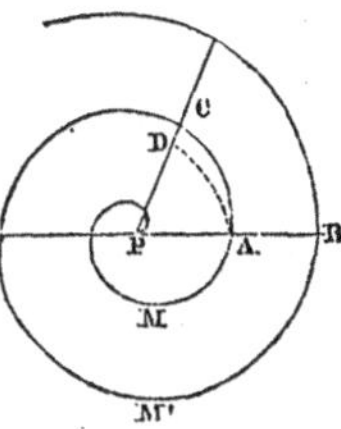

The system of polar co-ordinates is generally used to determine the different points of a spiral, and its equation may be represented by

$$r = f(v),$$

in which r denotes the radius vector, and v the variable angle.

138. Before discussing the particular spirals, it will be necessary to determine general expressions for the subtangent, &c., and the differentials of the arc and area, in terms of polar co-ordinates.

The subtangent, in such case, is the part of the perpendicular to the radius vector of the point of contact, intercepted between the pole and the point where the tangent meets this perpendicular. Thus, if A be the pole, and MT the tangent, AT perpendicular to AM is the subtangent. To find the expression for it, let the arc s receive the increment PP′ (AP being $= 1$); describe MC with the radius AM $= r$; draw the chords MC and MM′, and the line AT′ parallel to MC, and produce MM′ to T′. From the similar triangles MM′C and M′AT′, we have

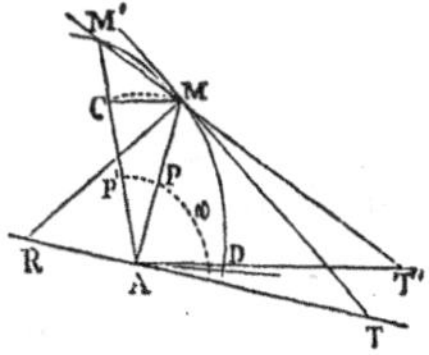

$$\text{M}'\text{C} : \text{MC} :: \text{AM}' : \text{AT}'; \qquad \text{AT}' = \frac{\text{MC} \times \text{AM}'}{\text{M}'\text{C}} \ldots (1)$$

Also, from the similar sectors APP′ and AMC,

$$1 : \text{PP}' :: \text{AM} : \text{arc MC}; \qquad \text{arc MC} = \text{AM} \times \text{PP}'.$$

Now, suppose the increment PP′ $= dv$, then M′C $= dr$, Art. (88), M′ becomes consecutive with M, the secant M′T′ coincides with the tangent MT, AT′ = AT, AM′ = AM = r, and chord MC = arc MC = rdv.

Making these substitutions in (1), we have

$$\text{AT} = \textit{subtangent} = \frac{r^2 dv}{dr} \dots\dots\dots\dots (2).$$

From this we deduce

$$\frac{\text{AT}}{r} = \frac{\text{AT}}{\text{AM}} = \frac{rdv}{dr} = \tan \text{AMT}.$$

$$\textit{The tangent } \text{MT} = \sqrt{\overline{\text{AM}}^2 + \overline{\text{AT}}^2} = r\sqrt{1 + r^2\frac{dv^2}{dr^2}}.$$

The similar triangles AMT and AMR, give

$$\text{AT} : r :: r : \text{AR}; \qquad \text{AR} = \frac{r^2}{\text{AT}} = \frac{dr}{dv} = \textit{subnormal}.$$

When M′ is consecutive with M, MM′C may be regarded as a triangle, right-angled at C; hence,

$$\text{MM}' = \sqrt{\overline{\text{M}'\text{C}}^2 + \overline{\text{MC}}^2}.$$

But MM′ is the differential of the arc; therefore

$$dz = \sqrt{dr^2 + r^2 dv^2}.$$

If ADM be any segment, AMM′ will be its increment when v is increased by dv. Calling the segment s, AMM′ will then be ds, and may be measured by the sector AMC. But the aıea of the sector,

$$AMC = \frac{1}{2}AM \times \text{arc } MC = \frac{r^2dv}{2};$$

hence,

$$ds = \frac{r^2dv}{2}.$$

139. An equation from which the particular equations of most of the spirals may be deduced, by assigning particular values to a and n, is

$$r = av^n.$$

If n be positive, $v = 0$ will give $r = 0$,

and the spirals represented by the equation have their origin at the pole.

If n be negative, $v = 0$ will give $r = \infty$,

and the spirals have their origin at an infinite distance, continually approach the pole, and r becomes equal to 0 only when $v = \infty$.

140. Let $n = 1$, then $r = av$,

and if r' and v', r'' and v'', represent the co-ordinates of any two points of the spiral, we shall have

$$r' = av', \qquad r'' = av'';$$

whence

$$r' : r'' :: v' : v'',$$

or the law in accordance with which the generating point must move is, *that the radius vectors shall be proportional to the corresponding angles.*

The curve thus generated is the *Spiral of Archimedes.*

If we take for the unit of distance, the length of the radius vector after one revolution; then $r = 1$, $v = 2\pi$, and the equation gives

$$1 = a.2\pi, \qquad a = \frac{1}{2\pi},$$

and the primitive equation becomes

$$r = \frac{v}{2\pi}; \qquad \text{whence} \qquad dr = \frac{dv}{2\pi}.$$

This spiral may be constructed by dividing a circumference into any number of equal parts, as 8, and the radius AB into the same number of equal parts. On the radius AC lay off one of these parts; on AD two, AE three, &c.; on AB eight, then again on AC nine, &c. The distances thus laid off will be proportional to the angles BAC, BAD, &c., and the curve through their extremities the required spiral.

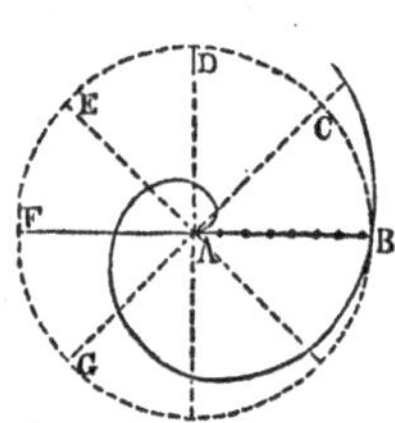

Substituting the values of r and dr in equation (2), Art. (138), we have

$$\text{AT} = \textit{subtangent} = \frac{v^2}{2\pi}.$$

If $v = 2\pi$, that is, if the tangent be drawn at the extremity of the arc generated in one revolution, we have

$$\text{AT} = 2\pi = \textit{circumference of measuring circle.}$$

If $v = m.2\pi$, or the tangent be drawn at the extremity of the arc generated in m revolutions,

$$\text{AT} = m^2.2\pi = m.2m\pi;$$

that is, equal to *m times the circumference described with the radius vector of the point of contact.*

For the subnormal we find

$$AR = \frac{dr}{dv} = \frac{1}{2\pi}.$$

141. If $n = \frac{1}{2}$, the general equation becomes

$$r = av^{\frac{1}{2}}, \qquad \text{or} \qquad r^2 = a^2v.$$

This equation being of the same form as that of the parabola, the curve given by it is called the *Parabolic Spiral.*

It may be constructed by first constructing the parabola whose equation is $y^2 = a^2x$, and then laying off from P to B, C, D, &c., along the circumference, any assumed abscissas, and from A to M, M′, &c., the corresponding ordinates; the points M, M′, &c., will be points of the spiral, since for each we have

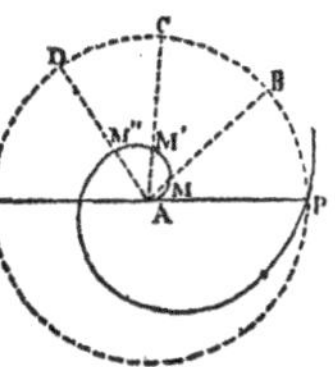

$$y^2 = a^2x, \qquad \text{or} \qquad r^2 = a^2v.$$

The subtangent at any point is $\qquad AT = \frac{2r^3}{a^2}.$

142. If $n = -1$, $r = av^n$ becomes

$$r = av^{-1} = \frac{a}{v}, \qquad \text{or} \qquad rv = a,$$

and the spiral thus given is called the *Hyperbolic Spiral.*

If r' and v', r' and v'', be the co-ordinates of any two points of the spiral, we have $r' = \frac{a}{v'}$, and $r'' = \frac{a}{v''}$; whence

$$r' : r'' :: \frac{1}{v'} : \frac{1}{v''},$$

or *the radius vectors are inversely proportional to the angles.*

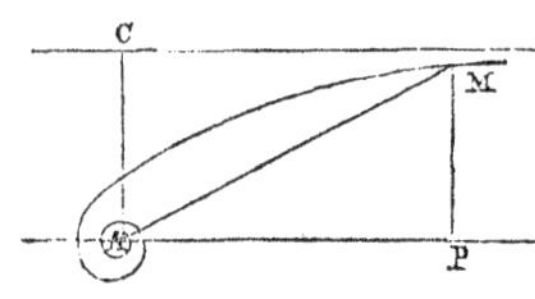

If M be any point of the spiral,

$$AM = r, \qquad MAP = v.$$

The right-angled triangle MAP, gives

$$r = \frac{MP}{\sin v}.$$

Substituting this value of r in the equation $rv = a$, we find

$$MP = a\frac{\sin v}{v}.$$

As v is diminished, this value approaches nearer to a, and since $\left(\frac{\sin v}{v}\right)_{v=0} = 1$, when $v = 0$, we have $MP = a$.

If then, at a distance, $AC = a$, a line be drawn parallel to AP, it will continually approach the curve, and touch it at an infinite distance.

$$\textit{The subtangent } AT = \frac{r^2 dv}{dr} = -a.$$

It is then constant, and equal to AC. Also,

$$\frac{rdv}{dr} = \tan AMT = -v;$$

that is, *the tangent of the angle made by the tangent and radius vector, is equal to the arc which measures the angle made by the radius vector and fixed line.*

We may apply these properties to the construction of the curve by points, thus: With A as a centre and radius $= a$, describe a circle; join any point T with A, draw the indefinite radius vector AM perpendicular to AT. Make AD = arc PN; join D and N, and draw TM parallel to DN, M will be a point of the curve; for by the construction

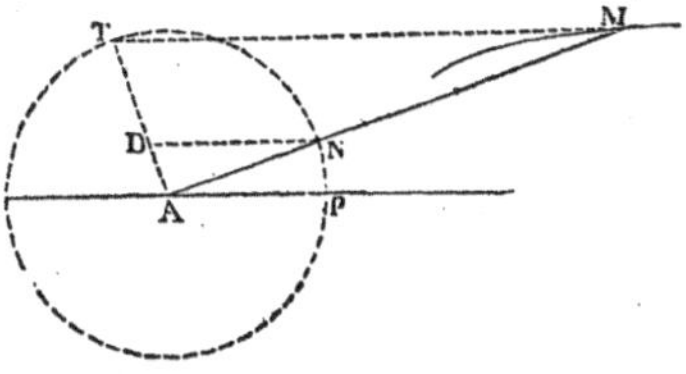

$$\text{AD} = \tan \text{AND} = \tan \text{AMT} = \text{arc NP}.$$

143. The spiral represented by the equation

$$v = \log r,$$

is called the *Logarithmic Spiral.*

Differentiating, we find

$$dv = \frac{\text{M}dr}{r};$$

whence

$$\tan \text{AMT} = \frac{rdv}{dr} = \text{M};$$

that is, *the angle formed by the radius vector and tangent is constant,* and the tangent of this angle is equal to the modulus of the system of logarithms used.

If the Naperian system be chosen, $\text{M} = 1$, and $\text{AMT} = 45°$

Since v is the logarithm of r, if it be increased uniformly, so that the different arcs v, v', v'', &c., shall be in arithmetical progression, then r, r', r'', &c., must be in geometrical progression, and the curve may be constructed thus: With AO $= 1$ describe a circle, and divide the circumference into any number of equal parts, and draw the lines AO, Ap, Ap', &c. The distances laid off on these lines are to be in geometrical progression, since the arcs Op, Op', Op'', &c., increase by the constant difference Op. To find the ratio of this progression, let $v = 0$, then $r =$ AO $= 1$. Now make $v =$ the arc Op, and find the corresponding value of r in the system of logarithms used, which lay off to m, then

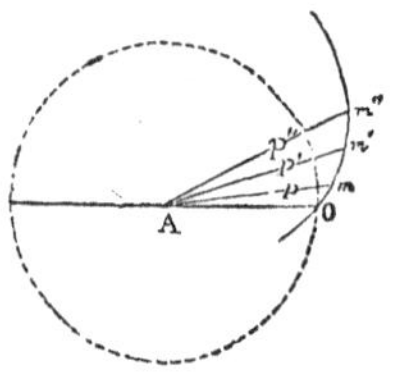

$$\frac{\mathrm{A}m}{\mathrm{AO}} = \text{the ratio.}$$

On Ap', Ap'', &c., lay off Am', Am'', so that

$$\mathrm{AO} : \mathrm{A}m : \mathrm{A}m' : \mathrm{A}m'' : \mathrm{A}m''' : \text{\&c.},$$

m, m', m'', &c., will be points of the curve.

Application of the Calculus to Surfaces.

144. Since the equation of every surface expresses the relation between the co-ordinates of its points, it must contain three variables, and may be generally written

$$u = \mathrm{F}(x, y, z) = 0 \ldots\ldots\ldots (1);$$

or since either two of these variables may be assumed at pleasure, and the remaining one determined from the equation, the latter may be regarded as a function of the other two, they being entirely independent of each other, and the equation of the surface be thus otherwise expressed,

$$z = f(x, y) \ldots\ldots\ldots\ldots (2).$$

By the same course of reasoning as that in Art. (79), it may be proved that every function of two variables may be regarded as the ordinate of a surface of which the variables are abscissas.

In the equation of every surface considered, z will be regarded as a function of x and y; and the co-ordinate planes will be taken at right angles to each other.

The differential equation of a surface may then be obtained, either by differentiating equation (1), as in article (57), or by differentiating equation (2), as in article (52). By the latter method, we obtain

$$dz = \frac{dz}{dx}dx + \frac{dz}{dy}dy \ldots\ldots (3).$$

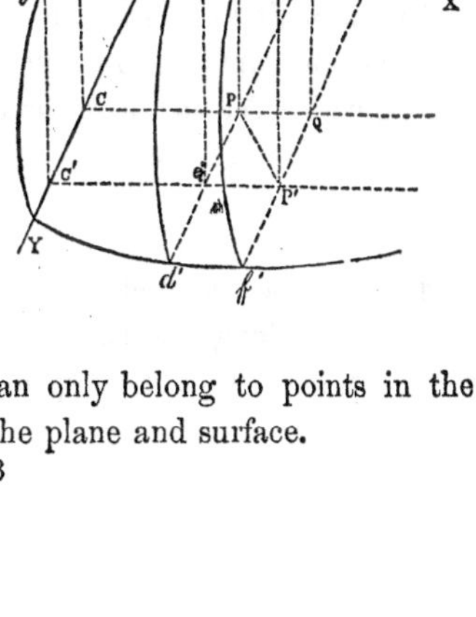

145. Let M be any point of a surface, a portion of which is represented in the annexed figure. The co-ordinates of this point are

$$x = \text{A}b, \quad y = \text{A}c, \quad z = \text{MP}.$$

Let a plane be passed through M, parallel to YZ. For every point of this plane,

$$x = \text{A}b = x''.$$

If, then, in the equation of the surface, we make $x = x''$, and suppose z and y to vary, they can only belong to points in the curve $d\text{M}d'$, the intersection of the plane and surface.

In the same way, if $y = y''$, in the equation of the surface, and z and x vary, we shall have the curve eMN.

If x and y vary at the same time, and receive the increments $bb' = h$ and $cc' = k$, we have

$$\text{M}'\text{P}' = z' = f(x + h,\ y + k),$$

which may be developed as in Art. (48).

When $x = x''$, equation (3), Art. (144), gives

$$dz = \frac{dz}{dy}dy = p'dy \qquad \text{or} \qquad \frac{dz}{dy} = p' \ldots\ldots (4);$$

equations which evidently belong only to the section dMd' parallel to YZ.

If $y = y''$, the corresponding equations for the section parallel to XZ are

$$dz = \frac{dz}{dx}dx = pdx, \qquad \text{or} \qquad \frac{dz}{dx} = p \ldots\ldots (5).$$

The value of $\frac{dz}{dy}$, equation (4), is the tangent of the angle which a tangent to the section dMd', at any point, makes with the axis of Y, or with the plane XY; and $\frac{dz}{dx}$, equation (5), the corresponding expression for the section eMN; and since these angles are the same as those made by the curves, at the point of contact, with XY, they give the inclination or *slope* of the surface in the direction of these curves.

146. If it be required to find the slope of the surface at any point, as M, along the section MM′ made by the plane MM′PP′, we take the equation of this plane,

$$y = \alpha x + \beta \ldots . (1), \qquad z \text{ indeterminate};$$

α being the tangent of the angle made with the axis of X by the trace PP′, and equal to $\frac{dy}{dx} = \frac{k}{h}$.

Now, in order that z shall represent only the ordinates of points in the section MM′, the relation expressed in equation (1) must exist between the variables x and y, and we must have

$$dy = \alpha dx,$$

which, in equation (3) of article (144), gives

$$dz = (p + \alpha p') dx.$$

The limit of the ratio $\frac{\text{M'P'} - \text{MP}}{\text{PP'}}$ is evidently the tangent of the angle (S) which the tangent, and consequently the curve at the point M, makes with PP′, or with the plane XY.

But since

$$\text{PP'} = \sqrt{\overline{\text{P'Q}}^2 + \overline{\text{PQ}}^2} = h\sqrt{1 + \alpha^2},$$

we have

$$\frac{\text{M'P'} - \text{MP}}{\text{PP'}} = \frac{z' - z}{h\sqrt{1 + \alpha^2}},$$

the limit of which is

$$\frac{1}{\sqrt{1 + \alpha^2}} \times \frac{dz}{dx} = \frac{p + \alpha p'}{\sqrt{1 + \alpha^2}} = \tan \text{S}.$$

To find the direction in which the section MM′ must be made, in order that the slope at a given point M, along the curve cut out, be greater than along any other, it is only necessary to obtain that value of α which will render the expression

$$\frac{p + \alpha p'}{\sqrt{1 + \alpha^2}}$$

a maximum, the values of p and p' being taken at the given point M. Differentiating the expression with reference to α, and placing the result equal to 0, we have

$$\frac{p' - p\alpha}{(1 + \alpha^2)^{\frac{3}{2}}} = 0;$$

whence

$$p' - p\alpha = 0, \qquad \alpha = \frac{p'}{p}.$$

This value of α substituted in equation (1), (β being first determined by the condition that the line PP′ shall pass through P), will give an equation which, combined with that of the surface, will determine the line of *greatest slope.*

Equations of Tangent Plane and Normal Line.

147. The co-ordinates of a given point M, being x'', y'', and z'', the equations of a tangent to the section parallel to XZ at this point, will be

$$z - z'' = \frac{dz''}{dx''}(x - x''), \qquad y = y'' \ldots .(1);$$

and to the section parallel to YZ,

$$z - z'' = \frac{dz''}{dy''}(y - y''), \qquad x = x'' \ldots .(2).$$

The equations of a plane passing through the same point, Analyt. Geom., Art. (64), will be

$$z - z'' = c(x - x'') + d(y - y'') \ldots\ldots(3).$$

The intersection of this plane by the plane through M parallel to XZ, will be represented by

$$z - z'' = c(x - x''), \qquad y = y'';$$

and the intersection by the plane parallel to YZ, by

$$z - z'' = d(y - y''), \qquad x = x''.$$

If the plane (3) is tangent to the surface at M, these lines should be tangent to the sections of the surface, and therefore identical with those represented by equations (1) and (2), and we must have

$$c = \frac{dz''}{dx''}, \qquad d = \frac{dz''}{dy''};$$

and equation (3) becomes *the equation of the tangent plane*,

$$z - z'' = \frac{dz''}{dx''}(x - x'') + \frac{dz''}{dy''}(y - y'') \ldots\ldots (4).$$

To illustrate, take the equation of the ellipsoid

$$\frac{x^2}{a^2} + \frac{y^2}{b^2} + \frac{z^2}{c^2} = 1,$$

from which, by differentiating, first with reference to x, and then with reference to y, and substituting z'', x'', and y'', we obtain

$$\frac{dz''}{dx''} = -\frac{c^2x''}{a^2z''} \qquad \frac{dz''}{dy''} = -\frac{c^2y''}{b^2z''}.$$

These expressions in equation (4) give, after reduction,

$$\frac{xx''}{a^2} + \frac{yy''}{b^2} + \frac{zz''}{c^2} = 1.$$

148. The equations of a straight line passing through the point M, are

$$x - x'' = a(z - z''), \qquad y - y'' = b(z - z'').$$

This will become perpendicular to the tangent plane, if we have the conditions, Analyt. Geom., Art. (59),

$$a = -c, \ b = -d, \quad \text{or} \quad a = -\frac{dz''}{dx''}, \ b = -\frac{dz''}{dy''},$$

and we thus deduce the equations of *a normal line*,

$$x - x'' = -\frac{dz''}{dx''}(z - z''), \qquad y - y'' = -\frac{dz''}{dy''}(z - z'').$$

By substituting these expressions for $x - x''$ and $y - y''$ in the general expression

$$D = \sqrt{(x - x'')^2 + (y - y'')^2 + (z - z'')^2},$$

we obtain for the distance from any point of the normal to the point of contact,

$$D = (z - z'')\sqrt{1 + \left(\frac{dz''}{dx''}\right)^2 + \left(\frac{dz''}{dy''}\right)^2}.$$

If $z = 0$,

$$D = z''\sqrt{1 + \left(\frac{dz''}{dx''}\right)^2 + \left(\frac{dz''}{dy''}\right)^2},$$

for the distance from the point where the normal pierces the plane XY to the point of contact, the minus sign being omitted, as the numerical value only is required. z'' divided by this distance, gives the sine of the angle which the normal makes with the plane

XY; and this angle is the complement of the angle made by the tangent plane with the plane XY; hence, we have, denoting this angle by β,

$$\cos\beta = \frac{1}{\sqrt{1 + \left(\frac{dz''}{dx''}\right)^2 + \left(\frac{dz''}{dy''}\right)^2}}.$$

Partial Differentials of a Surface and Volume.

149. Let BMM′ be any curve in space, and B′PP′ its projection on the co-ordinate plane XY. Let the plane of the curve MM′ make an angle β with the plane XY, and let its intersection with that plane be taken for the axis of X. Then, if the ordinate of the curve be denoted by y, the ordinate of its projection by y', the area of the curve by s, and that of its projection by s', we have, Art. (92),

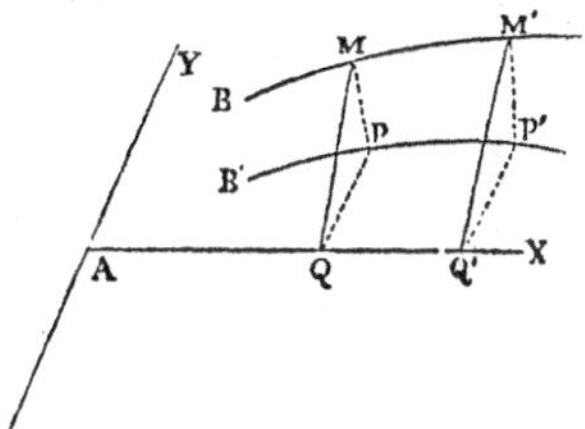

$$ds = y\,dx; \qquad ds' = y'\,dx.$$

The right-angled triangle MPQ gives $y' = y\cos\beta$; hence,

$$ds' = \cos\beta\, y\,dx = \cos\beta\, ds,$$

and the sum of all the values of ds' is equal to the sum of all the values of ds multiplied by $\cos\beta$. But the sum of all the values of ds' is the area s', Art. (88), and the sum of all the values of ds is the area s; hence,

$$s' = \cos\beta\, s;$$

that is, *the projection of any plane area is equal to the area multiplied by the cosine of the angle included between its plane and the plane of projection.*

150. Now, let u denote the area of any surface, as $ZeMd$, and M any point of the surface, whose co-ordinates are x, y, and z.

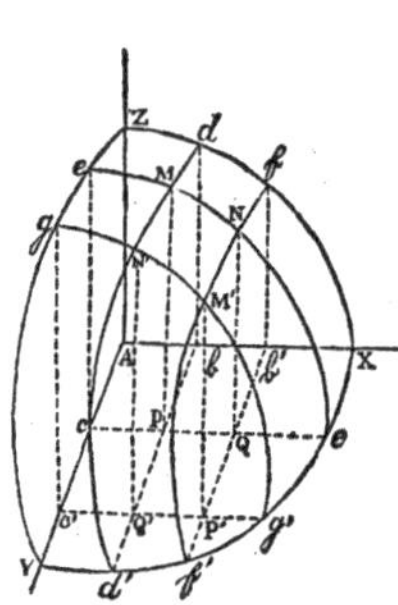

Since the equation of the surface gives z in terms of x and y, the area u is manifestly a function of x and y. Let x be increased by $bb' = dx$, y remaining the same, the increment of the surface will be $MdfN$, which will be the partial differential of u taken with respect to x; that is

$$MdfN = \frac{du}{dx}dx \ldots\ldots\ldots (1).$$

If now in this, y be increased by $PQ' = dy$, and x remain the same, the increment $MNM'N'$ will be the partial differential of (1) taken with respect to y; that is,

$$MNM'N' = \frac{d^2u}{dx\,dy}dx\,dy.$$

The same result may be obtained by first increasing y and then x.

The infinitely small area $MNM'N'$ may be regarded as a plane area in the tangent plane at M, and will, by the preceding article, be equal to the area of its projection $PQP'Q' = dx\,dy$, divided by $\cos\beta$. β being the angle made by the tangent plane with XY; hence

$$\frac{d^2u}{dx\,dy}dx\,dy = \frac{dx\,dy}{\cos\beta} = dx\,dy\sqrt{1 + \left(\frac{dz}{dx}\right)^2 + \left(\frac{dz}{dy}\right)^2} \ldots \text{Art. } (148),$$

or $$d^2u = dx\,dy \sqrt{1 + \left(\frac{dz}{dx}\right)^2 + \left(\frac{dz}{dy}\right)^2} \ldots\ldots(2),$$

in which it should be remembered that d^2u is the partial differential of the second order, obtained by differentiating first with respect to one variable, and then with resoect to the other.

151. Let v represent any volume limited by a surface, and the co-ordinate planes as AbPc-MZ. It will be a function of x and y. If x and y be increased in succession by dx and dy, as in the preceding article, we obtain first the increment volume

$$bb'\text{QP-N}d = \frac{dv}{dx}dx,$$

and for the increment of this, the volume

$$\text{PQP}'\text{Q}'\text{-M}'\text{M} = \frac{d^2v}{dx\,dy}dx\,dy.$$

But this infinitely small volume does not differ from the parallelopipedon whose base is $\text{PQP}'\text{Q}' = dx\,dy$, and altitude $\text{MP} = z$; hence

$$\frac{d^2v}{dx\,dy}dx\,dy = z\,dx\,dy, \quad \text{or} \quad d^2v = z\,dx\,dy \ldots.(1);$$

in which d^2v is a partial differential.

152. One surface is osculatory to another, when it has with it a more intimate contact than any other surface of the same kind; and the conditions which must exist in order that a surface, given in kind only, shall be osculatory to a given surface at a given point, can be determined by a method similar to that pursued in article (95). But from the nature of the case, these conditions are more numerous and complicated, and their determination more difficult; so much so as to render osculatory surfaces of little use in the measure of curvature; hence another method has been devised which will now be explained.

Let M be any point of a surface, at which it is proposed to examine the curvature. Let this point be taken as the origin of co-ordinates, and let the normal at this point be the axis of Z, the axes of X and Y having any position in the tangent plane XMY. The equation of the surface, Art. (144), will be

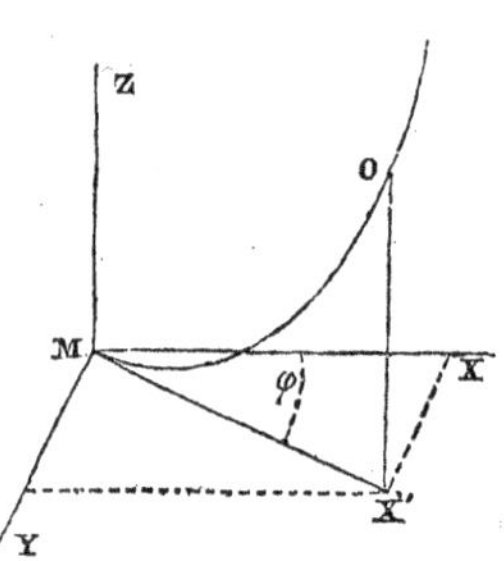

$$z = f(x, y) \ldots\ldots\ldots (1).$$

Through the normal let any plane ZMX', making an angle φ with the plane ZX, be passed; it will cut from the surface a curve MO. For any point of this curve, as O, denoting the abscissa MX' by x', we shall have

$$x = x' \cos \varphi, \qquad y = x' \sin \varphi \ldots\ldots\ldots (2),$$

and these values, substituted in equation (1), will evidently give the equation of the curve referred to the two axes MZ and MX'. Now, by varying the angle φ, all the normal sections at the point M may be obtained, and by examining the curvatures of these different sections at the given point, an accurate idea of the curvature of the surface may be formed.

Differentiating equations (2), we have

$$dx = dx' \cos \varphi, \qquad dy = dx' \sin \varphi \ldots\ldots (3).$$

The general expression for the radius of curvature of one of the normal sections, Art. (106), is

$$R = \frac{(dx'^2 + dz^2)^{\frac{3}{2}}}{dx' d^2 z} \dots\dots\dots\dots\dots (4).$$

Differentiating equation (3), Art. (144), we have, Art. (53),

$$d^2 z = \frac{d^2 z}{dx^2} dx^2 + 2 \frac{d^2 z}{dx dy} dx dy + \frac{d^2 z}{dy^2} dy^2 \dots\dots (5).$$

Substituting the above values of dx and dy in equation (3), Art. (144), and in (5), we have

$$dz = (p \cos\varphi + p' \sin\varphi) dx' \dots\dots\dots\dots (6),$$

$$d^2 z = (q \cos^2\varphi + 2q' \cos\varphi \sin\varphi + q'' \sin^2\varphi) dx'^2 \dots (6');$$

p and p' representing the partial differential coefficients of the first, and q, q', and q'' those of the second order of the function z.

If these values of dz and $d^2 z$ be substituted in expression (4), we shall have the expression for the radius of curvature of any one of the normal sections. But as we only desire this for the point M, we may first substitute the co-ordinates of this point, which are

$$x'' = 0, \qquad y'' = 0, \qquad z'' = 0;$$

and since the normal at this point coincides with the axis of Z, we must also have, Art. (148),

$$\frac{dz''}{dx''} = 0, \quad \frac{dz''}{dy''} = 0, \qquad \text{or} \qquad p = 0, \quad p' = 0.$$

Substituting these values in equations (6) and (6′), and the results in equation (4), we obtain

$$R = \frac{1}{q \cos^2\varphi + 2q' \cos\varphi \sin\varphi + q'' \sin^2\varphi} \dots (7);$$

in which q, q', and q'' are what the partial differential coefficients of the second order of the function z become, when 0 is substituted for x, y, and z.

Dividing by $\cos^2\varphi$, and recollecting that $\frac{1}{\cos^2\varphi} = 1 + \tan^2\varphi$, this value may be put under the form

$$R = \frac{1 + \tan^2\varphi}{q + 2q'\tan\varphi + q''\tan^2\varphi} \ldots\ldots\ldots (8).$$

We have taken the positive value of R, Art. (106), since, as the surface is represented in the figure, the sections are above the axis of X′, and convex towards it; $\frac{d^2z}{dx'^2}$ must therefore be positive, Art. (86), and the value of R positive, as it should be when laid off from M above the plane XY. If the section at the point M lies below the plane XY, it must still be convex towards this tangent plane; $\frac{d^2z}{dx'^2}$ will be negative, and R negative, and must therefore be laid off from M below XY.

By assigning all values to φ from 0 to 360° in equation (8), we shall obtain a value of R for each normal section. Among these values there must be one which is grèater, and another which is less than all the others. The values of φ which will give these principal values of R, will be obtained as in Art. (69).

Differentiating equation (8), we have

$$\frac{dR}{d\tan\varphi} = \frac{2(q'\tan^2\varphi + (q - q'')\tan\varphi - q')}{(q + 2q'\tan\varphi + q''\tan^2\varphi)^2}.$$

If the denominator be placed equal to 0, we shall obtain values of the tan φ which, when real, will reduce the value of R to infinity. The curvature of the corresponding section will then be zero, and the section itself a right line, or the point M a singular point, Art. (118), cases which do not occur in all surfaces. Let us then place the numerator equal to 0; we thus have

$$\tan^2\varphi + \frac{q - q''}{q'}\tan\varphi - 1 = 0 \ldots\ldots.(9).$$

This being either of the first or second form of equations of the second degree, the roots will always be real, and their product equal to -1, that is, denoting them by $\tan\varphi'$ and $\tan\varphi''$,

$$\tan\varphi' \tan\varphi'' + 1 = 0;$$

hence, the normal planes in which the greatest and least radii of curvature are found, must be perpendicular to each other. The sections by these planes are called *principal sections*, and their exact position will be determined by solving equation (9).

The values of $\tan\varphi'$ and $\tan\varphi''$ being determined, and the traces of the normal planes constructed as in the figure; let us take MX″ as a new axis of X, and MY″ as a new axis of Y, and suppose the surface to be referred to them, with MZ as an axis of Z. Then we must have for these new axes,

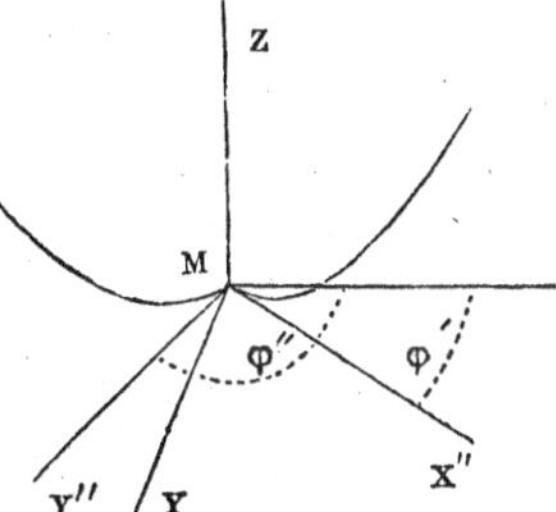

$$\tan\varphi' = 0, \qquad \tan\varphi'' = \infty, \qquad \tan\varphi' + \tan\varphi'' = \infty,$$

which requires in equation (9), that $q' = 0$. Substituting this value of q' in equation (7), we have

$$R = \frac{1}{q\cos^2\varphi + q''\sin^2\varphi} \ldots\ldots\ldots(10).$$

Substituting in this the values of φ, corresponding to the maximum and minimum radii as above determined, viz., $\varphi = 0$ and $\varphi = 90°$, and denoting the values of the principal radii thus determined by R′ and R″, we have

$$R' = \frac{1}{q}, \qquad R'' = \frac{1}{q''},$$

and finally, from equation (10),

$$\frac{1}{R} = q\cos^2\varphi + q''\sin^2\varphi = \frac{1}{R'}\cos^2\varphi + \frac{1}{R''}\sin^2\varphi,$$

which expresses the reciprocal of the radius of curvature of any normal section, in terms of the principal radii and the angle φ.

If R' and R'' are both positive, all values of R will be positive, and the greatest of the two will be a maximum, and the least a minimum; and all the normal sections at the point M will lie above the plane XY.

If R' and R'' are both negative, the sections will lie below XY. If one is positive and the other negative, a part of the values of R will be positive and a part negative, and a part of the sections will be above and a part below the plane XY; and this plane will cut the surface at the point M, giving a point analogous to the point of inflexion, Art. (92).

If $R' = R''$, all the values of R become equal to R' or R'', and the curvature of all the sections will be the same; as at any point of a sphere, or at the vertex of a surface of revolution.

153. If R''' be a value of R, in any section perpendicular to the one which makes the angle φ, we may obtain its reciprocal by substituting in the last expression for $\frac{1}{R}$, $90° + \varphi$ for φ; and since

$$\sin(90° + \varphi) = \cos\varphi, \qquad \cos(90° + \varphi) = -\sin\varphi,$$

this expression will become

$$\frac{1}{R'''} = \frac{1}{R'}\sin^2\varphi + \frac{1}{R''}\cos^2\varphi.$$

Adding this to the expression for $\frac{1}{R}$, member to member, we have

$$\frac{1}{R} + \frac{1}{R'''} = \frac{1}{R'} + \frac{1}{R''};$$

that is, Art. (105), *the sum of the curvatures of any two normal sections at the same point, which are perpendicular to each other, is constant, and equal to the sum of the curvatures of the principal sections.*

PART II.

INTEGRAL CALCULUS.

First Principles.

154. We have seen that the sum of all the values of a differential is the function from which it is derived, Art. (88). By whatever process this function may be obtained from its differential, it amounts to *the summation of the infinite number of its elements*, or infinitely small values of the differential. This process is called *integration*, and its symbol is $\int$, which always indicates an operation the reverse of differentiation; thus

$$\int du = u.$$

The object of the Integral Calculus is to explain how to pass from differentials to the functions from which they may be derived; or in any particular case, *to find an expression which, if it be differentiated, will produce the given differential.*

This expression is called *the integral* of the differential.

155. We have found, article (17), $d\text{A}u = \text{A}du$; therefore

$$\int \text{A}du = \int d\text{A}u = \text{A}u = \text{A}\int du.$$

From which we see that *a constant factor* may be placed without the sign of integration, without affecting the value of the integral; thus,

$$\int b(a - x^2)\,dx = b\int(a - x_2)\,dx, \qquad \int\frac{x^3dx}{c} = \frac{1}{c}\int x^3dx.$$

Also, in article (20), we have

$$d(u + v \pm \text{\&c.}) = du + dv \pm \text{\&c.};$$

hence

$$\int(du + dv \pm \text{\&c.}) = \int d(u + v \pm \text{\&c.}) = u + v \pm \text{\&c.}$$

$$= \int du + \int dv \pm \text{\&c.};$$

that is, *the integral of the sum or difference of any number of differentials, is equal to the sum or difference of their respective integrals.*

Also, in article (16), we have

$$d(u + C) = du,$$

no matter what the value of the constant C may be; hence an infinite number of expressions differing from each other in a constant term, when differentiated will produce the same differential. For this reason, *to complete the integral immediately found, we add a constant;* thus,

$$\int du = u + C.$$

Integration of Monomial Differentials.

156. By article (24), we have

$$cdx^{m+1} = c(m+1)x^m dx;$$

and from this,

$$cx^m dx = \frac{cdx^{m+1}}{m+1} = cd\frac{x^{m+1}}{m+1};$$

hence

$$\int cx^m dx = \int cd\frac{x^{m+1}}{m+1} = \frac{cx^{m+1}}{m+1} + C.$$

Therefore, to obtain the integral of a monomial differential: *Multiply the variable, with its primitive exponent increased by unity, by the constant factor, if there is one, and divide the result by the new exponent.*

Examples.

1. If $\quad du = xdx, \qquad \int du = \int xdx = \frac{x^2}{2} + C.$

2. If $\quad du = \frac{x^3 dx}{c}, \qquad \int du = \frac{1}{c}\int x^3 dx = \frac{x^4}{4c} + C.$

3. If $\quad du = bx^{\frac{2}{3}}dx, \qquad u = \frac{bx^{\frac{5}{3}}}{\frac{5}{3}} = \frac{3bx^{\frac{5}{3}}}{5} + C.$

4. If $\quad du = \frac{x^{-\frac{m}{n}}dx}{e}, \qquad u = \frac{nx^{\frac{n-m}{n}}}{e(n-m)} + C.$

5. If $$du = \frac{adx}{\sqrt{x}} + \frac{3x^{-\frac{2}{3}}dx}{b} - \frac{c^2x^4dx}{e},$$

$$u = \int\frac{adx}{\sqrt{x}} + \int\frac{3x^{-\frac{2}{3}}dx}{b} - \int\frac{c^2x^4dx}{e}\ldots\text{Art. (155)}.$$

The application of the above rule does not give the proper integral when $m = -1$, as in this case we have

$$\int x^{-1}dx = \frac{x^{-1+1}}{-1+1} = \frac{1}{0} = \infty;$$

whereas

$$\int x^{-1}dx = \int\frac{dx}{x} = lx + C\ldots\ldots\ldots\text{Art. (37)}.$$

This result was to be expected, since $\int\frac{dx}{x}$ or lx cannot be expressed in algebraic terms, Art. (5).

If $$du = \frac{a}{b}\frac{dx}{x},$$

$$u = \frac{a}{b}\int\frac{dx}{x} = \frac{a}{b}lx + C,$$

or $$u = \log x + C,$$

the logarithm being taken in the system whose modulus is $\frac{a}{b}$.

157. If we have an expression of the form

$$du = (a + bx + cx^2 + \&c.)^m x^n dx,$$

in which m is a positive whole number, the integral may be found by raising the quantity within the parenthesis to the mth power, multiplying each term by $x^n dx$, and then integrating it as in the preceding article.

Examples.

1. Let $$du = (a + x^2)^2 x^3 dx,$$

or
$$du = (a^2 + 2ax^2 + x^4)x^3 dx;$$

then
$$u = \int(a^2x^3dx + 2ax^5dx + x^7dx) = \frac{a^2x^4}{4} + \frac{2ax^6}{6} + \frac{x^8}{8} + C.$$

2. Let $$du = (b - x^2)^3 x^{\frac{1}{2}} dx.$$

3. Let $$du = (b - cx^{\frac{1}{2}})^2 x^{-\frac{2}{3}} dx.$$

Integration of Binomial Differentials of Particular Forms.

158. Many expressions, by the introduction of an auxiliary variable, may be transformed into monomials, and then integrated as in the preceding article.

I. Let $$du = (a + bx^n)^m c' x^{n-1} dx.$$

Place $$a + bx^n = z,$$

then
$$nbx^{n-1}dx = dz, \qquad x^{n-1}dx = \frac{dz}{bn}.$$

Substituting in the given expression, and integrating, we have

$$\int du = \int \frac{c'z^m dz}{bn} = \frac{c'}{bn}\int z^m dz = \frac{c'}{bn}\frac{z^{m+1}}{m+1};$$

and replacing the value of z, we have, finally,

$$u = \frac{c'(a+bx^n)^{m+1}}{(m+1)nb} + C;$$

that is, to integrate a binomial differential when the exponent of the variable without the parenthesis is one less than that within: *Multiply the binomial, with its primitive exponent increased by unity, by the constant factor, if there is one; then divide this result by the product of the new exponent, the coefficient, and the exponent of the variable within the parenthesis.*

Examples.

1. If $du = (a + bx^2)^{\frac{3}{2}} exdx$, $\qquad u = \dfrac{e(a+bx^2)^{\frac{5}{2}}}{\frac{5}{2}\cdot b\cdot 2} + C.$

2. If $du = (2 - 3x^5)^{-\frac{1}{2}} 3x^4 dx$, $\quad u = -\dfrac{2}{5}(2 - 3x^5)^{\frac{1}{2}} + C.$

3. If $du = (a - bx^{\frac{2}{3}})^{-\frac{3}{2}} x^{-\frac{1}{3}} dx$, $\quad u = \dfrac{3(a - bx^{\frac{2}{3}})^{-\frac{1}{2}}}{b} + C.$

4. Let $\qquad du = a(b - cz^{-\frac{m}{n}})^{-\frac{p}{q}} z^{-\frac{m}{n}-1} dz.$

II. Let $\qquad du = \dfrac{ax^{n-1}dx}{b \pm x^n}.$

Place $$b \pm x^n = z;$$

then

$$\pm nx^{n-1}dx = dz, \qquad x^{n-1}dx = \pm \frac{dz}{n},$$

and

$$u = \pm \int \frac{adz}{nz} = \pm \frac{a}{n} lz = \pm \frac{a}{n} l(b \pm x^n) + C.$$

In the same way, we may find the integrals of the following expressions:

1. Let $$du = \frac{m(b + 2cx)dx}{a + bx + cx^2}.$$

Place $a + bx + cx^2 = z$, then $(b + 2cx)dx = dz$,

$$u = m\int \frac{dz}{z} = mlz = ml(a + bx + cx^2) + C.$$

2. If $du = \frac{2dy}{a - y}$, $\qquad u = -2l(a - y) + C.$

3. If $du = \frac{(2 + 2x)dx}{2x + x^2}$, $\qquad u = l(2x + x^2) + C.$

4. Let $$du = \frac{2z^{\frac{1}{2}}dz}{1 - z^{\frac{3}{2}}}.$$

Since, in general,

$$\int \frac{adu}{u} = alu,$$

we see that in all cases where the numerator of an expression is the product of a constant and the differential of the denominator, *its integral will be the product of the constant and the Naperian logarithm of the denominator.*

159. Every expression of the form

$$du = \mathrm{A}x^m(a + bx)^n dx,$$

can be integrated, when either m or n is a positive whole number.

If n be positive and entire, we may integrate as in article (157).

If m be positive and entire, n being either fractional or negative, place

$$a + bx = z, \qquad \text{then} \qquad x = \frac{z - a}{b},$$

$$dx = \frac{dz}{b}, \qquad du = \mathrm{A}\left(\frac{z - a}{b}\right)^m z^n \frac{dz}{b},$$

$$u = \frac{\mathrm{A}}{b}\int\left(\frac{z - a}{b}\right)^m z^n dz,$$

which may be integrated as in article (157). The value of z being then replaced, the integral will be expressed in terms of x.

Examples.

1. Let $$du = bx^2(a - x)^{\frac{1}{2}}dx.$$

Place $a - x = z$, then $x = a - z$, $dx = -dz$,

$$u = \int - b(a - z)^2 z^{\frac{1}{2}}dz = -\frac{2}{3}ba^2z^{\frac{3}{2}} + \frac{4}{5}baz^{\frac{5}{2}} - \frac{2}{7}bz^{\frac{7}{2}},$$

and finally, by replacing the value of z,

$$u = -\frac{2}{3}ba^2(a - x)^{\frac{3}{2}} + \frac{4}{5}ba(a - x)^{\frac{5}{2}} - \frac{2}{7}b(a - x)^{\frac{7}{2}} + \mathrm{C}$$

2. If
$$du = \frac{2x dx}{(1 - 3x)^{\frac{1}{2}}},$$

it may be placed under the form

$$du = 2x(1 - 3x)^{-\frac{1}{2}}dx; \quad \text{whence} \quad u = -\frac{2}{9}\int(1 - z)z^{-\frac{1}{2}}dz,$$

and finally,

$$u = -\frac{4}{9}(1 - 3x)^{\frac{1}{2}} + \frac{4}{27}(1 - 3x)^{\frac{3}{2}} + C.$$

3. Let $du = -\dfrac{x^3 dx}{1 - x}$. 4. Let $du = \dfrac{y dy}{(3 - 2y)^{\frac{1}{2}}}$.

If
$$du = \frac{(Ax^m + Bx^p + Cx^q + \&c.)}{(ax + b)^n}dx,$$

we may place it under the form

$$du = \frac{Ax^m dx}{(ax + b)^n} + \frac{Bx^p dx}{(ax + b)^n} + \&c.,$$

and may then integrate each fraction as above, if m, p, q, &c., are entire and positive.

Use of the Arbitrary Constant. Integration between Limits.

160. To complete each integral as determined by the preceding rules, we have added a constant quantity C. If, in the particular case under consideration, we happen to know what the integral must be for a particular value of the variable, this constant can be determined. Thus, if

$$\int X dx = X' + C \ldots\ldots\ldots (1),$$

X' representing the function of x, obtained at once by the application of the rules for integration, and we know the integral must reduce to N when $x = a$, we have

$$N = X'_{x=a} + C, \qquad C = N - X'_{x=a}.$$

In general, however, this constant is entirely arbitrary, since whatever value be assigned to it, it will disappear by differentiation, Art. (16). This arbitrary nature of the constant enables us to cause the integral to fulfil any reasonable condition. Thus if, in equation (1), it be required that the integral reduce to the particular expression M, when $x = a$; we may determine the value which must be assigned to C, by writing M for $\int X dx$, and substituting a for x in the function X'. Calling the result of this substitution A, the equation reduces to

$$M = A + C; \qquad \text{whence} \qquad C = M - A,$$

and

$$\int X dx = X' + M - A \ldots\ldots (2),$$

which will fulfil the required condition.

$$\text{If} \quad M = 0, \quad C = -A \quad \text{and} \quad \int X dx = X' - A.$$

The integral $\int X dx = X' + C$, before any particular value has been assigned to C, is called *a complete, or indefinite* integral, and expresses the indefinite sum of all the values of the differential. After a particular value has been assigned to C, as in equation (2), it is called *a particular integral*, and expresses the sum of all values of the differential, commencing at *the origin of the integral*, that is, at that particular value of the integral which is 0. If, in this particular integral, a particular value be given to x, the result is called *a definite integral*. We should thus have, when $x = b$,

$$\int X dx = B + M - A \ldots\ldots (3),$$

B representing $X'_{x=b}$; and this expresses the definite sum of all values of the differential, from the value at the origin to that which corresponds to $x = b$.

That value of the variable which causes the integral to reduce to 0, and belongs to the origin, is always found by placing the particular integral equal to 0, and solving the resulting equation.

If in (1) we make $x = a$, and then $x = b$, we have

$$\int (X dx)_{x=a} = A + C, \qquad \int (X dx)_{x=b} = B + C,$$

whence, by subtraction,

$$\int (X dx)_{x=b} - \int (X dx)_{x=a} = B - A.$$

This is *the integral taken between the limits a and b*, and is usually written

$$\int_a^b X dx = B - A,$$

the limit corresponding to the subtractive integral being placed below. This expresses the definite sum of all values of the differential, between those which correspond to $x = a$ and $x = b$.

This *integral between limits* may always be obtained from either the complete or particular integral, by substituting, in succession, those values of the variable which indicate the limits, and subtracting the results.

If $a, b, c \ldots\ldots k, l$, be several increasing values of x, and we have

$$\int_a^b X dx = A', \quad \int_b^c X dx = B', \ldots\ldots \int_k^l X dx = K';$$

then evidently

$$\int_a^l X dx = A' + B' + C' \ldots\ldots + K'.$$

Example.

$$\int 6x^2dx = 2x^3 + C,$$

is a complete or indefinite integral.

If it be required that this reduce to 4, when $x = 1$, we have

$$4 = 2 + C, \qquad C = 2,$$

and

$$\int 6x^2dx = 2x^3 + 2,$$

the particular integral.

For the integral between the limits $x = 0$ and $x = 3$,

$$\int (6x^2dx)_{x=0} = 2, \qquad \int (6x^2dx)_{x=3} = 56;$$

hence,

$$\int_0^3 6x^2dx = 54.$$

The value of x corresponding to the origin of the particular integral, is obtained by placing

$$2x^3 + 2 = 0; \qquad \text{whence} \qquad x^3 = -1, \qquad x = -1.$$

Integration of the Differentials of the Simple Circular Functions, and of Circular Arcs.

161. By a reference to article (43), we see that

1. $\int \cos x\,dx = \sin x.$
2. $\int -\sin x\,dx = \cos x.$
3. $\int \sin x\,dx = \text{ver-sin}\, x.$
4. $\int \frac{dx}{\cos^2 x} = \tan x.$
5. $\int -\frac{dx}{\sin^2 x} = \cot x.$
6. $\int \tan x \,.\, \sec x\,dx = \sec x$, &c.

and from these we readily derive the integrals of the expressions,

7. $$\int 2x \cos x^2 dx = \int \cos x^2 . 2x dx = \sin x^2.$$

8. $$\int -\frac{1}{x^2} \sin \frac{1}{x} dx = \int - \sin \frac{1}{x} . \frac{dx}{x^2} = - \cos \frac{1}{x}.$$

9. $$\int - \frac{2x dx}{\cos^2(a - x^2)} = \int \frac{d(a - x^2)}{\cos^2(a - x^2)} = \tan(a - x^2).$$

These integrals will be completed by adding to each a constant. Hereafter, as in these examples, this constant will be understood, and written only when it is necessary for the discussion of the integral.

162. By a reference to article (44), we see that

1. $$\int \frac{du}{\sqrt{1 - u^2}} = \sin^{-1} u.$$ 2. $$\int \frac{-du}{\sqrt{1 - u^2}} = \cos^{-1} u.$$

3. $$\int \frac{du}{\sqrt{2u - u^2}} = \text{ver-sin}^{-1} u.$$ 4. $$\int \frac{du}{1 + u^2} = \tan^{-1} u.$$

And from these we derive the integrals of the similar expressions,

5. $$\int \frac{du}{\sqrt{a^2 - u^2}} = \int \frac{\frac{du}{a}}{\sqrt{1 - \frac{u^2}{a^2}}} = \int \frac{d\frac{u}{a}}{\sqrt{1 - \frac{u^2}{a^2}}} = \sin^{-1} \frac{u}{a}.$$

6. $$\int \frac{3 dx}{\sqrt{2 - x^2}} = 3 \int \frac{dx}{\sqrt{2 - x^2}} = 3 \sin^{-1} \frac{x}{\sqrt{2}}.$$

7. $$\int \frac{2\,dx}{\sqrt{9 - 3x^2}} = \int \frac{2}{\sqrt{3}} \frac{dx}{\sqrt{3 - x^2}} = \frac{2}{\sqrt{3}} \sin^{-1} \frac{x}{\sqrt{3}}.$$

8. $$\int \frac{-\,du}{\sqrt{a^2 - u^2}} = \int \frac{-\frac{du}{a}}{\sqrt{1 - \frac{u^2}{a^2}}} = \cos^{-1} \frac{u}{a}.$$

9. $$\int - \frac{2\,du}{\sqrt{4 - u^2}} = 2\int - \frac{du}{\sqrt{4 - u^2}} = \cos^{-1} \frac{u}{2}.$$

10. $$\int \frac{du}{\sqrt{2au - u^2}} = \int \frac{\frac{du}{a}}{\sqrt{2\frac{u}{a} - \frac{u^2}{a^2}}} = \text{ver-sin}^{-1} \frac{u}{a}.$$

11. $$\int \frac{3\,dx}{\sqrt{4x - 2x^2}} = \frac{3}{\sqrt{2}} \int \frac{dx}{\sqrt{2x - x^2}} = \frac{3}{\sqrt{2}} \text{ver-sin}^{-1} x.$$

12. $$\int \frac{du}{a^2 + u^2} = \int \frac{\frac{du}{a^2}}{1 + \frac{u^2}{a^2}} = \frac{1}{a} \int \frac{\frac{du}{a}}{1 + \frac{u^2}{a^2}} = \frac{1}{a} \tan^{-1} \frac{u}{a}.$$

13. $$\int \frac{2\,dx}{2 + 3x^2} = \frac{2}{3} \int \frac{dx}{\frac{2}{3} + x^2} = \frac{2}{\sqrt{6}} \tan^{-1} \frac{x}{\sqrt{\frac{2}{3}}}.$$

INTEGRATION OF RATIONAL FRACTIONS.

163. Every rational fraction which is the differential of a function of x, will appear as a particular case of the general form

$$\frac{(Ax^m + Bx^{m-1} + Cx^{m-2} + \&c.)\,dx}{A'x^n + B'x^{n-1} + C'x^{n-2} + \&c.},$$

in which m and n are whole numbers, and positive.

If m be greater than n, the numerator may be divided by the denominator, and the division continued until the greatest exponent of x in the remainder is, at least, one less than in the denominator; the quotient will then consist of an entire and rational part, plus the remainder divided by the denominator, and may be written

$$\frac{(Ax^m + Bx^{m-1} + \&c.)\,dx}{A'x^n + B'x^{n-1} + \&c.} = Xdx + \frac{(A''x^{n-1} + B''x^{n-2} + \&c.)\,dx}{A'x^n + B'x^{n-1} + \&c.};$$

and the integral of the primitive fraction will be the sum of the integrals of the two parts.

It will be necessary, then, to explain only the manner of integrating the second part, or those rational fractions in which the greatest exponent of the variable in the numerator is at least one less than in the denominator.

First, suppose the denominator to be divided into its simple factors of the first degree, and let them be represented by

$$x - a, \quad x - b, \quad x - c, \quad \&c.$$

There will be four different cases, each of which will reqaire a separate discussion.

I. *When the factors are real and unequal;*
II. *When they are real and equal;*
III. *When they are imaginary, and no two alike;*
IV. *When they are imaginary and alike, two and two.*

164. I. As an example of the first case, let us take the fraction

$$\frac{(ax + c)\,dx}{x^2 - b^2}.$$

The two factors of the denominator are $x + b$ and $x - b$; then

$$\frac{(ax + c)\,dx}{x^2 - b^2} = \frac{(ax + c)\,dx}{(x + b)(x - b)}.$$

Place

$$\frac{ax + c}{x^2 - b^2} = \frac{\text{A}}{x + b} + \frac{\text{A}'}{x - b}\ldots\ldots(1),$$

A and A′ being constants to be determined. For the purpose of determining them, clear the equation of its denominators; then

$$ax + c = \text{A}x - \text{A}b + \text{A}'x + \text{A}'b.$$

Since this is true for all values of x, by the principle of indeterminate coefficients, we may place the coefficients of the like powers of x, in the two members, equal to each other, and have

$$a = \text{A} + \text{A}', \qquad c = \text{A}'b - \text{A}b,$$

$$\text{A} = \frac{ab - c}{2b}, \qquad \text{A}' = \frac{ab + c}{2b}.$$

Substituting these values in (1), multiplying by dx, and prefixing the sign $\int$, we have

$$\int\frac{(ax + c)\,dx}{x^2 - b^2} = \frac{ab - c}{2b}\int\frac{dx}{x + b} + \frac{ab + c}{2b}\int\frac{dx}{x - b}$$

$$= \frac{ab - c}{2b}l(x + b) + \frac{ab + c}{2b}l(x - b).$$

If there be n factors in the denominator of the given expression, there should be n corresponding fractions of the above form; and these, when reduced to a common denominator with the first member, will give an identical equation, containing n different powers of the variable (including the zero power), from which, as above, n equations may be formed and the values of the n numerators be determined.

The method pursued above indicates the following rule for all similar expressions:

Place the primitive fraction (omitting the differential of the variable) equal to the sum of as many partial fractions as there are factors of the first degree in its denominator; the numerators of these fractions being constants to be determined, and the denominators the several factors of the original denominator; clear the resulting equation of denominators, equate the coefficients of the like powers of the variable in the two members, and thence determine the constants; then multiply each partial fraction by the differential of the variable, and take the sum of their integrals as in case II., *article* (158).

2. Integrate the expression

$$\frac{(3x^2 - 1)}{x^3 - x}dx.$$

The factors of the denominator are, $x + 1$, $x - 1$, and x; hen

$$\frac{3x^2 - 1}{x^3 - x} = \frac{A}{x+1} + \frac{A'}{x-1} + \frac{A''}{x}.$$

Clearing of denominators,

$$3x^2 - 1 = Ax^2 - Ax + A'x^2 + A'x + A''x^2 - A'';$$

whence

$$3 = A + A' + A'', \qquad 0 = -A + A', \qquad 1 = A''$$

$$A = 1 = A' = A'',$$

and

$$\int \frac{(3x^2 - 1)}{x^3 - x} dx = \int \frac{dx}{x+1} + \int \frac{dx}{x-1} + \int \frac{dx}{x}$$

$$= l(x+1) + l(x-1) + lx = l(x^3 - x),$$

as may be seen at once, since the numerator of the given differential is the exact differential of the denominator.

3. Integrate the expression

$$\frac{(1-y)\,dy}{y^2 - 2y - 2}.$$

Placing the denominator equal to 0, we have

$$y^2 - 2y - 2 = 0;$$

whence $y = 1 \pm \sqrt{3}$, and the corresponding factors are

$y - (1 + \sqrt{3})$, $y - (1 - \sqrt{3})$, or $y - m$ and $y - n$.

Finally,

$$\int \frac{(1-y)dy}{y^2-2y-2} = \frac{m-1}{n-m}l(y-m) - \frac{n-1}{n-m}l(y-n).$$

4. Integrate

$$\frac{(2x+3)\,dx}{x^3 - x^2 - 2x}.$$

Integrate

$$\frac{(x^3-1)\,dx}{x^2-4}.$$

165. II. In the second case it may be remarked, that if all the factors of the denominator are equal, the fraction will take the form

$$\frac{Ax^{n-1} + Bx^{n-2} + \text{\&c.}}{(x-a)^n}dx,$$

which may be integrated as in article (159).

We need, then, only consider the case where a portion of the factors are equal. The rule of the preceding article is not applicable here, as will be seen by taking the expression

$$\frac{adx}{(x-b)^2(x-c)},$$

in which two of the factors are equal to $x - b$.

By an application of the rule referred to, we should have

$$\frac{a}{(x-b)^2(x-c)} = \frac{A}{x-b} + \frac{A'}{x-b} + \frac{A''}{x-c}$$

$$= \frac{A+A'}{x-b} + \frac{A''}{x-c} = \frac{B}{x-b} + \frac{A''}{x-c},$$

since $A + A'$ must be regarded as a single constant.

If this equation be cleared of denominators, and the coefficients of the like powers of x in the two members placed equal to each other, we shall evidently form three independent equations, with only two unknown quantities, B and A''.

We obviate this difficulty by writing, for the equal factors, the two fractions $\frac{B}{(x-b)^2}$ $\frac{B'}{x-b}$, and thus have

$$\frac{a}{(x-b)^2(x-c)} = \frac{B}{(x-b)^2} + \frac{B'}{x-b} + \frac{A}{x-c},$$

which, being cleared of denominators, gives

$$a = B(x-c) + B'(x-b)(x-c) + A(x-b)^2;$$

whence

$$B' + A = 0, \quad B - B'c - B'b - 2Ab = 0,$$

$$B'bc - Bc + Ab^2 = a,$$

three equations with three unknown quantities, which can then be determined.

And in general, if there be n equal factors, we should write n partial fractions in the form

$$\frac{B}{(x-b)^n} + \frac{B'}{(x-b)^{n-1}} \cdots\cdots + \frac{B^{(n-1)\prime}}{x-b},$$

the numerators of which are constants, and the denominators the different powers of the equal factor, from the nth down to the first power. After B, B', &c., are determined, each partial fraction, being first multiplied by the differential of the variable, will be integrated as in article (158).

Examples.

1. Integrate $\dfrac{(2 + x)dx}{(x - 1)^2(x - 2)}$.

Place

$$\frac{2 + x}{(x - 1)^2(x - 2)} = \frac{B}{(x - 1)^2} + \frac{B'}{x - 1} + \frac{A}{x - 2}.$$

Clearing of denominators, and equating the coefficients of the like powers of x, we have

$$0 = B' + A, \quad 1 = B - 3B' - 2A, \quad 2 = -2B + 2B' + A,$$

$$B = -3, \quad B' = -4, \quad A = 4;$$

and finally

$$\int \frac{(2 + x)dx}{(x - 1)^2(x - 2)} = \frac{3}{x - 1} - 4l(x - 1) + 4l(x - 2).$$

2. Integrate $\dfrac{xdx}{x^3 - x^2 - x + 1}$.

If there are different sets of equal factors, partial fractions must be written for each set; thus,

$$\frac{2}{(x - 1)^2(x + 1)^2} = \frac{A}{(x - 1)^2} + \frac{A'}{x - 1} + \frac{B}{(x + 1)^2} + \frac{B'}{x + 1}.$$

166. III. We know, from the general theory of equations, that imaginary roots are found only in pairs, and that for each pair we must have a factor of the second degree, which, placed equal to 0, will give the imaginary roots. Each pair of roots will always appear as a particular case of the general form

$$x = a \pm \sqrt{-b^2} \ldots\ldots\ldots (1),$$

and the corresponding factor of the second degree will be

$$x^2 - 2ax + a^2 + b^2 = [x - (a + \sqrt{-b^2})][x - (a - \sqrt{-}$$

By a comparison of the imaginary factors, in any given with these general expressions, we determine the correspond. values of a and b. Thus, if the factor of the second degree be

$$x^2 - 2x + 5,$$

we place it equal to 0, and find the two roots

$$x = 1 \pm \sqrt{-4};$$

whence, by comparison, $a = 1, \quad b^2 = 4, \quad b = 2.$

Now, in the third case, for each pair of imaginary factors, partial fraction be written, of the form

$$\frac{Mx + N}{x^2 - 2ax + a^2 + b^2} = \frac{Mx + N}{(x - a)^2 + b^2}.$$

By clearing of denominators, &c., as in the preceding articles, M and N may be determined. We shall have then to integrate the expression

$$\frac{(Mx + N)\,dx}{(x - a)^2 + b^2}.$$

For this purpose, make $x - a = z$, then $x = z + a$, $dx = dz$.

Substituting these, the original expression becomes

$$\frac{(Mz + Ma + N)}{z^2 + b^2}dz;$$

or, by making $\quad Ma + N = P, \quad$ and dividing the expression into two parts,

$$\frac{Mzdz}{z^2 + b^2} + \frac{Pdz}{z^2 + b^2}.$$

The first part may be integrated as in case II., Art. (158). Thus,

$$\int \frac{Mzdz}{z^2 + b^2} = \frac{M}{2} l(z^2 + b^2) = Ml\sqrt{z^2 + b^2} = Ml\sqrt{(x - a)^2 + b^2}.$$

The integral of the second part is

$$P\int \frac{dz}{z^2 + b^2} = \frac{P}{b}\tan^{-1}\frac{z}{b} \ldots\ldots \text{Art. (162)};$$

or, by substituting the values of P and z,

$$\int \frac{Pdz}{z^2 + b^2} = \frac{N + Ma}{b}\tan^{-1}\left(\frac{x - a}{b}\right);$$

and finally,

$$\int \frac{(Mx + N)\,dx}{(x - a)^2 + b^2} = Ml\sqrt{(x - a)^2 + b^2}$$

$$+ \frac{N + Ma}{b}\tan^{-1}\frac{(x - a)}{b} \ldots\ldots (2).$$

Take the particular example

$$\frac{(x - 1)\,dx}{x^3 + x^2 + 2x}.$$

The factors of the denominator are x and $x^2 + x + 2$, the latter being the product of the two factors corresponding to the imaginary roots

$$x = -\frac{1}{2} \pm \sqrt{-\frac{7}{4}},$$

which, compared with (1), give $a = -\frac{1}{2}$, $b^2 = \frac{7}{4}$, $b = \frac{1}{2}\sqrt{7}$

Place

$$\frac{x-1}{x^3+x^2+2x} = \frac{A}{x} + \frac{Mx+N}{x^2+x+2}.$$

Clearing of denominators, &c., we find

$$A = -\frac{1}{2}, \quad M = \frac{1}{2}, \quad N = \frac{3}{2}.$$

Substituting these values of M, N, a, and b, in formula (2), observing that $\int \frac{Adx}{x} = \int -\frac{1}{2}\frac{dx}{x} = -\frac{1}{2}lx$, and reducing, we have

$$\int \frac{(x-1)\,dx}{x^3+x^2+2x} = -\frac{1}{2}lx + \frac{1}{2}l\sqrt{x^2+x+2}$$

$$+ \frac{5}{2\sqrt{7}}\tan^{-1}\left(\frac{x+\frac{1}{2}}{\frac{1}{2}\sqrt{7}}\right).$$

167. IV. In the fourth case, where there are several imaginary factors, alike two and two, those of each pair multiplied together will give the same factor of the second degree; and if there be p such pairs, the denominator will contain a factor of the form

$$(x^2 - 2ax + a^2 + b^2)^p.$$

For this, we write p partial fractions; thus,

$$\frac{Mx+N}{[(x-a)^2+b^2]^p} + \frac{M'x+N'}{[(x-a)^2+b^2]^{p-1}} \cdots + \frac{M^{(p-1)'}x + N^{(p-1)'}}{(x-a)^2+b^2}.$$

Clearing of denominators, &c., the values of M, N, M′, N′, &c., may be determined as before; and since the several partial fractions, after multiplying by dx, are all of the same form, we have only to explain the mode of integrating any one of them, except the last, which is to be integrated as in the preceding article. Take the first,

$$\frac{(\mathrm{M}x + \mathrm{N})\,dx}{[(x-a)^2 + b^2]^p},$$

and make $x - a = z$; the fraction then becomes

$$\frac{(\mathrm{M}z + \mathrm{M}a + \mathrm{N})\,dz}{(z^2 + b^2)^p},$$

or, placing $\mathrm{M}a + \mathrm{N} = \mathrm{P}$,

$$\frac{\mathrm{M}z dz}{(z^2 + b^2)^p} + \frac{\mathrm{P}dz}{(z^2 + b^2)^p}.$$

The first part is integrated as in case I., Art. (158). Thus,

$$\int \frac{\mathrm{M}z dz}{(z^2 + b^2)^p} = \frac{\mathrm{M}(z^2 + b^2)^{-p+1}}{(-p+1)\,2} = \frac{\mathrm{M}}{2(1-p)(z^2 + b^2)^{p-1}}.$$

By means of a formula hereafter to be determined [Formula D, Art. (181)], we shall find

$$\int \frac{\mathrm{P}dz}{(z^2 + b^2)^p} = \varphi(z) + \mathrm{C}' \int \frac{dz}{z^2 + b^2} = \varphi(z) + \frac{\mathrm{C}'}{b} \tan^{-1}\frac{z}{b};$$

then

$$\int \frac{(\mathrm{M}z + \mathrm{P})\,dz}{(z^2 + b^2)^p} = \frac{\mathrm{M}}{2(1-p)(z^2 + b^2)^{p-1}} + \varphi(z) + \frac{\mathrm{C}'}{b} \tan^{-1}\frac{z}{b},$$

after which, substituting the value of z, we shall obtain the complete integral of the primitive expression.

168. By a review of the preceding discussion, it will be seen that all differentials which are rational fractions can be integrated, provided the factors of the denominator can be discovered; and that the integrals will depend upon one or more of the four forms,

$$\int \frac{dx}{x + a}, \quad \int x^m dx, \quad \int \frac{xdx}{(x^2 + a^2)^p}, \quad \int \frac{dx}{x^2 + a^2}.$$

Integration by Parts.

169. In article (21), we have found

$$duv = udv + vdu; \quad \text{whence} \quad uv = \int udv + \int vdu,$$

and

$$\int udv = uv - \int vdu \ldots\ldots (1);$$

from which we see that the integral of udv can be obtained, whenever we are able to integrate vdu. This method of integrating udv is called *integration by parts*. To apply it to a particular example, divide the given differential into two factors, one of which shall contain the differential of the variable, and be capable of immediate integration; substitute this factor for dv, its integral for v, and the other factor for u, in formula (1).

Examples.

1. Integrate the expression $\quad x^3 dx \sqrt{a - x^2}$.

This may be divided into the two factors,

$$x^2 \quad \text{and} \quad xdx\sqrt{a - x^2}.$$

Place $\quad x^2 = u \quad$ and $\quad xdx\sqrt{a - x^2} = dv;$

then

$$du = 2xdx, \qquad v = \int xdx\sqrt{a - x^2} = -\frac{(a - x^2)^{\frac{3}{2}}}{3}.$$

Substituting these in formula (1), we have

$$\int udv = -\frac{x^2(a - x^2)^{\frac{3}{2}}}{3} + \int\frac{(a - x^2)^{\frac{3}{2}}}{3}2xdx;$$

and finally,

$$\int x^3dx\sqrt{a - x^2} = -\frac{x^2(a - x^2)^{\frac{3}{2}}}{3} - \frac{2}{15}(a - x^2)^{\frac{5}{2}}$$

2. Integrate $\quad \dfrac{(1 - x^2)^{\frac{1}{2}}dx}{x^2}.$

Place $\quad (1 - x^2)^{\frac{1}{2}} = u, \quad$ and $\quad \dfrac{dx}{x^2} = dv;$

then

$$\int\frac{(1 - x^2)^{\frac{1}{2}}}{x^2}dx = -\frac{\sqrt{1 - x^2}}{x} - \sin^{-1}x.$$

3. Integrate $\quad \dfrac{x^2dx}{(a^2 - x^2)^{\frac{3}{2}}}.$

4. Integrate $\quad \dfrac{x^3dx}{\sqrt{a^2 - x^2}}.$

Integration of certain Irrational Differentials.

170. In the preceding articles, rules have been given *by which every rational differential may be integrated*, except the case referred to in article (168). It may then be taken for granted, that, in general, *every irrational differential which can be made rational in terms of a new variable, can also be integrated.* Let

$$\frac{ax^{\frac{h}{k}}\,dx}{bx^{\frac{m}{n}} + cx^{\frac{p}{q}}}$$

be a differential, all the *irrational parts* of which are *monomials.* Make

$$x = z^{knq}; \qquad \text{then} \qquad x^{\frac{h}{k}} = z^{hnq},$$

$$x^{\frac{m}{n}} = z^{mkq}, \qquad x^{\frac{p}{q}} = z^{knp}, \qquad dx = knqz^{knq-1}\,dz.$$

These values substituted in the given expression, evidently make it rational in terms of z and dz. It may then be integrated, after which the value of z in terms of x must be substituted. We may then enunciate the following rule for the integration of expressions of this kind: *For the variable substitute another, with an exponent equal to the least common multiple of the indices of the radicals; then integrate by the known rules, and substitute in the result the value of the new variable in terms of the primitive.*

Examples.

1. Let $$du = \frac{2x^{\frac{1}{2}} - 3x^{\frac{2}{3}}}{5x^{\frac{1}{6}}}dx\ldots\ldots(1).$$

The least common multiple of the denominators or indices being 6, we place

$$x = z^6, \qquad \text{then} \qquad dx = 6z^5dz, \qquad z = x^{\frac{1}{6}}.$$

Substituting in (1), we have

$$du = \frac{2z^3 - 3z^4}{5z} 6z^5dz = \frac{12}{5}z^7dz - \frac{18}{5}z^8dz,$$

and integrating,

$$u = \frac{12}{40}z^8 - \frac{18}{45}z^9 = \frac{3}{10}x^{\frac{4}{3}} - \frac{2}{5}x^{\frac{3}{2}}.$$

2. Let $du = \dfrac{3x^{\frac{1}{2}}dx}{2x^{\frac{1}{2}} - x^{\frac{2}{3}}}$. 3. Let $du = \dfrac{axdx}{b - c\sqrt{x}}$.

171. If the irrational parts are all of the form $(a + bx)^{\frac{p}{q}}$, the expression may be made rational in terms of z, by placing

$$a + bx = z^r,$$

r being the least common multiple of the indices of the radicals. We shall thus have

$$x = \frac{z^r - a}{b}, \qquad dx = \frac{rz^{r-1}dz}{b},$$

which, substituted in the primitive expression, with the value of $a + bx$, will evidently give a rational result. Take the examples:

1. $$du = \frac{dx}{(1 + x)^{\frac{3}{2}} + (1 + x)^{\frac{1}{2}}}.$$

Place $1 + x = z^2$; then $dx = 2zdz$, $z = (1 + x)^{\frac{1}{2}}$.

These values substituted in (1), give

$$du = \frac{2zdz}{z^3 + z} = \frac{2dz}{1 + z^2};$$

whence

$$u = 2\int \frac{dz}{1 + z^2} = 2\tan^{-1}z = 2\tan^{-1}(1 + x)^{\frac{1}{2}}.$$

2. Let $$du = \frac{dx}{x\sqrt{1 + x}}.$$

3. Let $$du = \frac{xdx}{(1 - x)^{\frac{1}{2}} + (1 - x)^{\frac{1}{4}}}.$$

172. Differentials of the form

$$Xdx\left(\frac{a + bx}{a' + b'x}\right)^{\frac{m}{n}}$$

X being a rational function of x, may be made rational by placing $\frac{a + bx}{a' + b'x} = z^n$, deducing the values of x and dx, and substituting them.

For example, let

$$du = xdx\left(\frac{1 + x}{1 - x}\right)^{\frac{2}{3}} \ldots\ldots.(1).$$

Place $\frac{1 + x}{1 - x} = z^3$, then $x = \frac{z^3 - 1}{z^3 + 1}$, $dx = \frac{6z^2dz}{(z^3 + 1)^2}$.

These values in (1), give

$$du = \frac{(z^3 - 1)\,6z^4dz}{(z^3 + 1)^3},$$

which is rational.

173. Every radical of the form $\sqrt{a + bx \pm cx^2}$. can be written thus

$$\sqrt{c}\sqrt{\frac{a}{c} + \frac{b}{c}x \pm x^2} = \sqrt{c}\sqrt{\alpha + \beta x \pm x^2},$$

after making $\frac{a}{c} = \alpha$, and $\frac{b}{c} = \beta$.

To render rational a differential, the only irrational part of which is a radical of the above form, it will then only be necessary to find rational values for x, dx, and $\sqrt{\alpha + \beta x \pm x^2}$, in terms of a new variable and its differential.

I. Take the case in which the sign of x^2 is $+$, and place

$$\sqrt{\alpha + \beta x + x^2} = z - x \ldots\ldots (1).$$

Squaring both members, we have

$$\alpha + \beta x = z^2 - 2zx;$$

whence

$$x = \frac{z^2 - \alpha}{\beta + 2z} \ldots\ldots\ldots\ldots (2).$$

By differentiating this value of x, we obtain

$$dx = \frac{2(z^2 + \beta z + \alpha)\,dz}{(\beta + 2z)^2} \ldots\ldots\ldots (3),$$

and by substituting the value of x in the second member of (1),

$$\sqrt{\alpha + \beta x + x^2} = \frac{z^2 + \beta z + \alpha}{\beta + 2z} \ldots (4).$$

These values of x, dx, and $\sqrt{\alpha + \beta x + x^2}$, substituted in the primitive differential, will evidently give a rational expression in terms of z and dz. After integrating this, the value of z, taken from (1), must be substituted.

Examples.

1. Let $du = \dfrac{dx}{\sqrt{a + bx + cx^2}} = \dfrac{dx}{\sqrt{c}\sqrt{\alpha + \beta x + x^2}}$.

Substituting for dx and $\sqrt{\alpha + \beta x + x^2}$ their values as found above, and reducing, we have

$$du = \frac{dx}{\sqrt{c}\sqrt{\alpha + \beta x + x^2}} = \frac{2\,dz}{\sqrt{c}(\beta + 2z)};$$

whence

$$u = \frac{1}{\sqrt{c}} l(\beta + 2z) = \frac{1}{\sqrt{c}} l[\beta + 2(\sqrt{\alpha + \beta x + x^2} + x)] \,..(5).$$

2 Let $du = \dfrac{dx}{\sqrt{h + c^2x^2}} = \dfrac{dx}{c\sqrt{\dfrac{h}{c^2} + x^2}}$.

By comparison with the similar expression in the preceding example, we see that

$$c = \sqrt{c}, \qquad 0 = \beta, \qquad \frac{h}{c^2} = \alpha.$$

Substituting these values in (5), we deduce

$$u = \int \frac{dx}{c\sqrt{\frac{h}{c^2} + x^2}} = \frac{1}{c} l 2 \left(\sqrt{\frac{h}{c^2} + x^2} + x \right)$$

$$= \frac{1}{c} l 2 \left(\frac{\sqrt{h + c^2x^2} + cx}{c} \right) = \frac{1}{c} l \frac{2}{c} + \frac{1}{c} l (\sqrt{h + c^2x^2} + cx);$$

and, finally, after uniting the constant $\frac{1}{c} l \frac{2}{c}$ with the arbitrary constant,

$$u = \int \frac{dx}{\sqrt{h + c^2x^2}} = \frac{1}{c} l (\sqrt{h + c^2x^2} + cx).$$

3. Let $$du = \frac{dx \sqrt{2x + x^2}}{x^2}.$$

Comparing this with formulas (2), (3), and (4), we see that

$$0 = \alpha, \qquad 2 = \beta, \qquad x = \frac{z^2}{2 + 2z};$$

$$dx = \frac{2(z^2 + 2z)\,dz}{(2 + 2z)^2}, \qquad \sqrt{2x + x^2} = \frac{z^2 + 2z}{2 + 2z};$$

whence

$$du = \frac{(z + 2)^2 dz}{z^2(z + 1)}.$$

4. Let $$du = dx \sqrt{m^2 + x^2}.$$

5. Let $$du = \frac{dx}{x \sqrt{x^2 - ax}}.$$

174. II. If the sign of x^2 be minus, it will be necessary to pursue a different method, and deduce other formulas; for if we write

$$\sqrt{\alpha + \beta x - x^2} = z - x,$$

the second powers of x in the squares of the two members will have contrary signs, and not cancel each other, as in the first case; and therefore the deduced value of x in terms of z will not be rational.

Denoting the roots of the equation $x^2 - \beta x - \alpha = 0$, by δ and δ', we have

$$x^2 - \beta x - \alpha = (x - \delta)(x - \delta');$$

or, changing the signs,

$$\alpha + \beta x - x^2 = (x - \delta)(\delta' - x), \quad \sqrt{\alpha + \beta x - x^2} = \sqrt{(x - \delta)(\delta' - x)}.$$

Now, if we make

$$\sqrt{(x - \delta)(\delta' - x)} = (x - \delta)z \ldots\ldots (1),$$

square both members, and strike out the common factor $x - \delta$, we have

$$\delta' - x = (x - \delta)z^2, \qquad x = \frac{\delta' + \delta z^2}{1 + z^2},$$

$$x - \delta = \frac{\delta' + \delta z^2}{1 + z^2} - \delta = \frac{\delta' - \delta}{1 + z^2} \ldots (2).$$

Substituting this value in equation (1), we obtain

$$\sqrt{\alpha + \beta x - x^2} = \sqrt{(x - \delta)(\delta' - x)} = \frac{(\delta' - \delta)}{1 + z^2}.$$

By differentiating equation (2), we find

$$dx = -\frac{2(\delta' - \delta)zdz}{(1 + z^2)^2}.$$

These values of x, $\sqrt{\alpha + \beta x - x^2}$, and dx, substituted in the primitive expression, will make it rational.

Examples.

1. Let $$du = \frac{dx}{\sqrt{\alpha + \beta x - x^2}}.$$

By substituting the values of dx and $\sqrt{\alpha + \beta x - x^2}$, we obtain

$$du = -\frac{2dz}{1 + z^2},$$

$$u = -2\int \frac{dz}{1 + z^2} = -2\tan^{-1} z;$$

and since, from equation (1),

$$z = \sqrt{\frac{\delta' - x}{x - \delta}},$$

we have, finally,

$$u = \int \frac{dx}{\sqrt{\alpha + \beta x - x^2}} = -2\tan^{-1}\sqrt{\frac{\delta' - x}{x - \delta}} + C.$$

If in this we make $\beta = 0$, $\alpha = 1$, the expression reduces to

$$u = \int \frac{dx}{\sqrt{1 - x^2}} = C - 2\tan^{-1}\sqrt{\frac{1 - x}{1 + x}};$$

since, by placing $x^2 - 1 = 0$, we find

$$x = \pm 1, \quad \text{or} \quad \delta = -1, \quad \delta' = 1.$$

If we now introduce the condition that the integral shall be 0, when $x = 0$, we have

$$0 = C - 2\tan^{-1}1 = C - \frac{\pi}{2}, \qquad C = \frac{\pi}{2},$$

and

$$\int \frac{dx}{\sqrt{1 - x^2}} = \frac{\pi}{2} - 2\tan^{-1}\sqrt{\frac{1 - x}{1 + x}}.$$

The direct integral of the first member is $\sin^{-1}x$, Art. (162); hence,

$$\sin^{-1}x = \frac{\pi}{2} - 2\tan^{-1}\sqrt{\frac{1 - x}{1 + x}}.$$

2. Let $$du = \frac{dx\sqrt{2ax - x^2}}{x^2}.$$

Placing $2ax - x^2 = 0$, we deduce $x = 0$, and $x = 2a$; hence, $\delta = 0$, and $\delta' = 2a$. Substituting these in the formulas, &c., we have

$$\frac{dx\sqrt{2ax - x^2}}{x^2} = -\frac{2z^2dz}{1 + z^2},$$

a simple rational fraction.

3. $du = \dfrac{xdx}{\sqrt{4x - x^2}}$. 4. $du = \dfrac{dx\sqrt{2 - x^2}}{x}$.

Integration of Binomial Differentials.

175. 1. If we have a differential of the form

$$x^{m-1}dx(ax^r + bx^n)^{\frac{p}{q}},$$

x^r (r being supposed less than n) may be taken out of the parenthesis, and for the primitive expression we may write

$$x^{m-1}dx\,x^{\frac{rp}{q}}(a + bx^{n-r})^{\frac{p}{q}} = x^{m+\frac{rp}{q}-1}dx(a + bx^{n-r})^{\frac{p}{q}},$$

in which but one of the terms, in the parenthesis, contains the variable x, and the exponent of this variable will be positive.

2. If, after this, the exponent of x should be fractional, either within or without the parenthesis, or both, we can substitute for x another variable, with an exponent equal to the least common multiple of the denominators of the given exponents, and thus get rid of the fractions; as in the example

$$x^{\frac{1}{3}}dx(a + bx^{\frac{1}{2}})^{\frac{p}{q}},$$

by making $x = z^6$, we obtain

$$x^{\frac{1}{3}}dx(a + bx^{\frac{1}{2}})^{\frac{p}{q}} = 6z^7dz(a + bz^3)^{\frac{p}{q}},$$

in which the exponents of z are whole numbers. *Hence, every binomial differential can be placed under the form*

$$x^{m-1}dx(a + bx^n)^{\frac{p}{q}},$$

in which m and n are whole numbers, and n positive.

176. 1. The binomial differential being placed under the proposed form, if $\frac{p}{q}$ is entire and positive, it may be integrated as in article (157); if $\frac{p}{q}$ is entire and negative, we have

$$x^{m-1}dx\,(a+bx^n)^{-\frac{p}{q}} = \frac{x^{m-1}dx}{(a+bx^n)^{\frac{p}{q}}},$$

which is a rational fraction.

2. If $\frac{p}{q}$ is a fraction, either positive or negative, place

$$a + bx^n = z^q;$$

then

$$(a+bx^n)^{\frac{p}{q}} = z^p \ldots.(1), \qquad x^n = \frac{z^q - a}{b},$$

$$x^m = \left(\frac{z^q - a}{b}\right)^{\frac{m}{n}}, \qquad x^{m-1}dx = \frac{1}{n}\left(\frac{z^q - a}{b}\right)^{\frac{m}{n}-1}\frac{qz^{q-1}}{b}dz\,..(2).$$

The values (1) and (2), substituted in the primitive expression, give

$$x^{m-1}dx\,(a+bx^n)^{\frac{p}{q}} = \frac{q}{nb}z^{p+q-1}dz\left(\frac{z^q - a}{b}\right)^{\frac{m}{n}-1} \ldots.(3),$$

which is rational in terms of z and dz, when $\frac{m}{n}$ *is a whole number.*

Example.

Let $$du = x^3 dx (a - bx^2)^{\frac{3}{2}},$$

in which

$$m - 1 = 3, \quad n = 2, \quad \frac{m}{n} = 2, \quad p = 3, \quad q = 2, \quad b = -b,$$

These values in equation (3), give

$$x^3 dx (a - bx^2)^{\frac{3}{2}} = z^4 dz \frac{(z^2 - a)}{b^2},$$

in which

$$z^2 = a - bx^2.$$

3. If $\frac{m}{n}$ is not a whole number, we may write

$$x^{m-1} dx (a + bx^n)^{\frac{p}{q}} = x^{m-1} dx \left[x^n \left(\frac{a}{x^n} + b\right)\right]^{\frac{p}{q}}$$

$$= x^{m+\frac{np}{q}-1} dx (ax^{-n} + b)^{\frac{p}{q}},$$

and, in accordance with the preceding principle, this may be made rational if

$$\frac{m + \frac{np}{q}}{-n} = -\left(\frac{m}{n} + \frac{p}{q}\right) \qquad \textit{is a whole number.}$$

To obtain the proper rational expression in terms of z, we need only make in equation (3),

$$m = m + \frac{np}{q}, \qquad n = -n, \qquad a = b, \qquad b = a.$$

Thus,

$$x^{m+\frac{np}{q}-1} dx (ax^{-n} + b)^{\frac{p}{q}} = -\frac{q}{na} z^{p+q-1} dz \left(\frac{z^q - b}{a}\right)^{-\frac{m}{n}-\frac{p}{q}-1} \ldots (4).$$

Example.

Let $$du = xdx(a + bx^3)^{\frac{1}{3}},$$

in which

$$m - 1 = 1, \quad n = 3, \quad p = 1, \quad q = 3, \quad \frac{m}{n} + \frac{p}{q} = 1$$

These values in equation (4) give

$$xdx(a + bx^3)^{\frac{1}{3}} = -\frac{z^3 dz}{a}\left(\frac{z^3 - b}{a}\right)^{-2},$$

in which $$z^3 = ax^{-3} + b.$$

From what precedes, we see that every binomial differential of the proposed form can be integrated, *if the exponent of the parenthesis is a whole number; if the exponent of the variable without the parenthesis, plus unity, divided by the exponent of the variable within, is a whole number;* or *if this quotient, plus the exponent of the parenthesis, is a whole number.*

177. Let us now write p for $\frac{p}{q}$, and then divide the expression

$$x^{m-1}dx(a + bx^n)^p = x^{m-n}x^{n-1}dx(a + bx^n)^p,$$

into the two factors

$$x^{m-n} = u, \qquad \text{and} \qquad x^{n-1}dx(a + bx^n)^p = dv;$$

whence

$$du = (m - n)x^{m-n-1}dx, \quad v = \frac{(a + bx^n)^{p+1}}{(p + 1)nb} \text{....Art. (158).}$$

Substituting these values in the formula

$$\int udv = uv - \int vdu \ldots\ldots \text{Art. (169)},$$

and making $(a + bx^n) = \text{X}$, we have

$$\int x^{m-1}dx\text{X}^p = \frac{x^{m-n}\text{X}^{p+1}}{(p+1)nb} - \frac{(m-n)}{(p+1)nb}\int x^{m-n-1}dx\text{X}^{p+1} \ldots (1)$$

But since

$$\text{X}^{p+1} = \text{X}^p\text{X} = \text{X}^p(a + bx^n) = a\text{X}^p + bx^n\text{X}^p,$$

$$\int x^{m-n-1}dx\text{X}^{p+1} = a\int x^{m-n-1}dx\text{X}^p + b\int x^{m-1}dx\text{X}^p.$$

Substituting this value in (1), and clearing of denominators,

$$(p + 1)nb\int x^{m-1}dx\text{X}^p$$

$$= x^{m-n}\text{X}^{p+1} - (m - n)\left[a\int x^{m-n-1}dx\text{X}^p + b\int x^{m-1}dx\text{X}^p\right];$$

transposing, &c., we obtain

$$\int x^{m-1}dx\text{X}^p = \frac{x^{m-n}\text{X}^{p+1} - a(m-n)\int x^{m-n-1}dx\text{X}^p}{b(pn + m)} \ldots \textbf{A}.$$

By a single application of this formula, we cause

$$\int x^{m-1}dx\text{X}^p \qquad \text{to depend upon} \qquad \int x^{m-n-1}dx\text{X}^p,$$

in which the exponent of the variable without the parenthesis is diminished by the exponent of the variable within. By an application of the same formula to $\int x^{m-n-1}dx\text{X}^p$, it may be made to depend upon $\int x^{m-2n-1}dx\text{X}^p$; and finally, by repeated applications, $\int x^{m-1}dx\text{X}^p$ will depend upon the expression

$$a(m - rn)\int x^{m-rn-1}dx\text{X}^p,$$

in which r represents the number of times m will contain n. If m is positive and an exact multiple of n, then $m - rn = 0$, the term containing the expression to be integrated disappears, and the integration is complete.

If $pn + m = 0$, the second member of the formula becomes infinite, and it fails to answer the purpose; but in this case $p + \frac{m}{n} = 0$, which, substituted in equation (4) of article (176), gives an expression which may at once be integrated.

178. We may also write

$$\int x^{m-1}dx\mathrm{X}^p = \int x^{m-1}dx\mathrm{X}^{p-1}\mathrm{X} = a\int x^{m-1}dx\mathrm{X}^{p-1} + b\int x^{m+n-1}dx\mathrm{X}^{p-1}.$$

If now in formula **A** we change m into $m + n$, and p into $p - 1$, we have

$$\int x^{m+n-1}dx\mathrm{X}^{p-1} = \frac{x^m\mathrm{X}^p - am\int x^{m-1}dx\mathrm{X}^{p-1}}{b\,(pn + m)}$$

Substituting this value in the preceding equation, and reducing, we obtain

$$\int x^{m-1}dx\mathrm{X}^p = \frac{x^m\mathrm{X}^p + pna\int x^{m-1}dx\mathrm{X}^{p-1}}{pn + m} \; \ldots\ldots \mathbf{B};$$

by which the primitive expression is made to depend upon another, in which the exponent of the parenthesis is one less than before. By repeated applications, this exponent, if positive, may be reduced to a fraction less than unity, either positive or negative.

179. The use of the preceding formulas may be illustrated by the example

$$\int x^2dx\,(a + bx^2)^{\frac{3}{2}}.$$

Place $a + bx^2 = X$, $m = 3$, $n = 2$, $p = \frac{3}{2}$; then, from formula **A**,

$$\int x^2 dx X^{\frac{3}{2}} = \frac{xX^{\frac{5}{2}} - a\int dx X^{\frac{3}{2}}}{6b}.$$

Applying formula **B** to the expression $\int dx X^{\frac{3}{2}}$, making $m = 1$ $n = 2$, $p = \frac{3}{2}$, we have

$$\int dx X^{\frac{3}{2}} = \frac{xX^{\frac{3}{2}} + 3a\int dx X^{\frac{1}{2}}}{4},$$

and by another application

$$\int dx X^{\frac{1}{2}} = \frac{xX^{\frac{1}{2}} + a\int \frac{dx}{X^{\frac{1}{2}}}}{2}.$$

Substituting these values, we have, finally,

$$\int x^2 dx X^{\frac{3}{2}} = \frac{xX^{\frac{5}{2}}}{6b} - \frac{axX^{\frac{3}{2}}}{24b} - \frac{a^2 xX^{\frac{1}{2}}}{16b} - \frac{a^3}{16b}\int \frac{dx}{X^{\frac{1}{2}}}.$$

The expression $\frac{dx}{X^{\frac{1}{2}}} = \frac{dx}{\sqrt{a + bx^2}}$ may be integrated as in Art. (173).

180. If, in the primitive expressions, m and p are negative, the effect of the application of formulas **A** and **B**, would evidently be to increase them numerically. Other formulas are then required.

From **A**, by transposition and reduction, we find

$$\int x^{m-n-1} dx X^p = \frac{x^{m-n} X^{p+1} - b(m + np)\int x^{m-1} dx X^p}{a(m - n)}.$$

If in this we change m into $-m+n$, we have

$$\int x^{-m-1}dxX^p = -\frac{x^{-m}X^{p+1} - b(n-m+np)\int x^{-m+n-1}dxX^p}{am} \text{..}\,\mathbf{C};$$

by the application of which, $-m$ will be numerically diminished by the number of units in n.

If $n - m + np = 0$, or $n = \dfrac{m}{p+1}$,

the part to be integrated will disappear, and the integration will be complete. If $m = 0$, the formula fails.

181. From **B**, by transposition and reduction, we find

$$\int x^{m-1}dxX^{p-1} = \frac{-x^mX^p + (m+np)\int x^{m-1}dxX^p}{pna}.$$

If in this we change p into $-p+1$, we obtain

$$\int x^{m-1}dxX^{-p} = \frac{x^mX^{-p+1} - (m+n-np)\int x^{m-1}dxX^{-p+1}}{an(p-1)} \text{....}\,\mathbf{D};$$

in which the exponent of X is numerically one less than in the primitive expression.

If $p - 1 = 0$, the second member becomes infinite; but in this case $p = 1$, and the primitive expression reduces to a rational fraction.

If $m + n - np = 0$, or $n = \dfrac{m}{p-1}$,

the simple application of the formula gives the integral at once.

182. Let us illustrate the use of these formulas by the example

$$\int x^{-2}dx\,(2 - x^2)^{-\frac{3}{2}}.$$

Making in **C**, $m = 1$, $a = 2$, $b = -1$, $n = 2$, $p = -\frac{3}{2}$, we have

$$\int x^{-2}dx\,(2 - x^2)^{-\frac{3}{2}} = -\frac{x^{-1}X^{-\frac{1}{2}}}{2} + \int dxX^{-\frac{3}{2}}\ldots(1).$$

By formula **D**, after making $m = 1$, $n = 2$, $a = 2$, $b = -1$, $p = \frac{3}{2}$, we have

$$\int dxX^{-\frac{3}{2}} = \frac{xX^{-\frac{1}{2}}}{2}.$$

Making the proper substitutions in (1), we obtain, finally,

$$\int x^{-2}dx\,(2 - x^2)^{-\frac{3}{2}} = -\frac{x^{-1}X^{-\frac{1}{2}}}{2} + \frac{xX^{-\frac{1}{2}}}{2}$$

in which $X = 2 - x^2$.

183. By the aid of formula **D**, we are now able to integrate the expression

$$\frac{dz}{(z^2 + b^2)^p} = dz\,(z^2 + b^2)^{-p}\ldots\text{Art. (167)}.$$

By making $m = 1$, $x = z$, $a = b^2$, $b = 1$, $n = 2$, we cause $\int\frac{dz}{(z^2 + b^2)^p}$ to depend upon the integration of another expression, in which the exponent is one less; and by repeated applications, we shall find that the integral will depend upon the expression

$$\int\frac{dz}{z^2 + b^2} = \frac{1}{b}\tan^{-1}\frac{z}{b}.$$

184. For the expression

$$\int \frac{x^q dx}{\sqrt{2cx - x^2}},$$

we may write

$$\int x^q dx \, (2cx - x^2)^{-\frac{1}{2}} = \int x^{q-\frac{1}{2}} dx \, (2c - x)^{-\frac{1}{2}},$$

to which applying formula **A**, making

$$m = q + \tfrac{1}{2}, \quad a = 2c, \quad b = -1, \quad p = -\tfrac{1}{2}, \quad n = 1,$$

and recollecting that $x^{q-\frac{1}{2}} = x^{q-1}x^{\frac{1}{2}}$, and $x^{q-\frac{3}{2}} = x^{q-1}x^{-\frac{1}{2}}$, we obtain

$$\int \frac{x^q dx}{\sqrt{2cx - x^2}} = -\frac{x^{q-1}\sqrt{2cx - x^2}}{q}$$

$$+ \frac{(2q-1)c}{q} \int \frac{x^{q-1} dx}{\sqrt{2cx - x^2}} \ldots\ldots \mathbf{E}.$$

By repeated applications of this formula, when q is a whole number, we make the primitive expression depend upon

$$\int \frac{dx}{\sqrt{2cx - x^2}} = \text{ver-sin}^{-1}\frac{x}{c} \ldots\ldots \text{Art. (161)}.$$

Integration by Series.

185. If it be required to integrate the expression Xdx, X being any function of x, it is often convenient and useful to develop X into a series by any of the known methods, generally by the binomial formula; and then, after multiplying by dx, to integrate each term separately. This is called *integrating by series;*

since we thus obtain a series equal to the integral of the given expression, from which, when the series is converging, we can, for particular values of the variable, deduce the approximate value of the integral.

1. Let us take the example

$$du = \frac{dx}{1 + x} = dx(1 + x)^{-1}.$$

By the binomial formula, we have

$$(1 + x)^{-1} = 1 - x + x^2 - x^3 + \&c.$$

Multiplying by dx, and prefixing the sign $\int$,

$$\int \frac{dx}{1 + x} = \int (dx - xdx + x^2dx - x^3dx + \&c.);$$

whence

$$\int \frac{dx}{1 + x} = x - \frac{x^2}{2} + \frac{x^3}{3} - \frac{x^4}{4} + \&c.;$$

or since

$$\int \frac{dx}{1 + x} = l(1 + x),$$

$$l(1 + x) = x - \frac{x^2}{2} + \frac{x^3}{3} - \frac{x^4}{4} + \&c. \ldots\ldots \text{Art. (38).}$$

2. Let $$du = x^{\frac{1}{2}}(1 - x^2)^{\frac{1}{2}}dx.$$

By the binomial formula, we have

$$(1 - x^2)^{\frac{1}{2}} = 1 - \frac{x^2}{2} - \frac{x^4}{8} - \frac{x^6}{16} - \&c.$$

Multiplying each term by $x^{\frac{1}{2}}dx$, &c.,

$$\int x^{\frac{1}{2}}(1 - x^2)^{\frac{1}{2}}dx = \int (x^{\frac{1}{2}}dx - \frac{x^{\frac{5}{2}}dx}{2} - \frac{x^{\frac{9}{2}}dx}{8} - \&c.);$$

whence

$$\int x^{\frac{1}{2}}(1 - x^2)^{\frac{1}{2}}dx = \frac{2}{3}x^{\frac{3}{2}} - \frac{1}{7}x^{\frac{7}{2}} - \frac{x^{\frac{11}{2}}}{44} - \&c.$$

3. Let $$du = a^x dx.$$

In article (40), we have found

$$a^x = 1 + la\frac{x}{1} + (la)^2\frac{x^2}{1.2} + (la)^3\frac{x^3}{1.2.3} + \&c.;$$

hence,

$$\int a^x dx = x + \frac{la\,x^2}{2} + \frac{(la)^2 x^3}{6} + \&c.$$

If $a = e$, then $le = 1$, and

$$\int e^x dx = x + \frac{x^2}{2} + \frac{x^3}{6} + \frac{x^4}{24} + \&c.$$

4. Let $$du = \frac{dx}{\sqrt{x - x^2}} = \frac{dx}{\sqrt{x}\sqrt{1 - x}}.$$

Make $\sqrt{x} = u$; then $dx = 2\sqrt{x}\,du$, and

$$\frac{dx}{\sqrt{x}\sqrt{1 - x}} = \frac{2du}{\sqrt{1 - u^2}},$$

which may be readily integrated, and we shall obtain

$$\int \frac{dx}{\sqrt{x - x^2}} = 2 \sin^{-1} \sqrt{x} = 2\sqrt{x}(1 + \frac{x}{2.3} + \frac{3x^2}{2.4.5} + \&c.....).$$

5. $du = e^{-x^2}dx.$ 6. $du = dx\sqrt{2ax - x^2}.$

7. Let $$du = \frac{dx\sqrt{1 - e'^2x^2}}{\sqrt{1 - x^2}}.$$

Developing $\sqrt{1 - e'^2x^2} = (1 - e'^2x^2)^{\frac{1}{2}}$, we have

$$\sqrt{1 - e'^2x^2} = 1 - \frac{1}{2}e'^2x^2 - \frac{1}{8}e'^4x^4 - \&c.;$$

hence

$$\int \frac{dx\sqrt{1 - e'^2x^2}}{\sqrt{1 - x^2}} = \int (1 - \frac{1}{2}e'^2x^2 - \frac{1}{8}e'^4x^4 - \&c.)\frac{dx}{\sqrt{1 - x^2}}.$$

After the multiplication, each term of the second member will be of the form $A\int \frac{x^m dx}{\sqrt{1 - x^2}}$, which, by formula **A**, may be made to depend upon

$$\int \frac{dx}{\sqrt{1 - x^2}} = \sin^{-1} x.$$

8. Let $du = \frac{dx}{\sqrt{(2cx - x^2)(b - x)}} = \frac{dx}{\sqrt{2cx - x^2}\sqrt{b - x}}.$

If we develop $\frac{1}{\sqrt{b - x}} = (b - x)^{-\frac{1}{2}}$, and multiply by $\frac{dx}{\sqrt{2cx - x^2}}$, each term will be of the form $\frac{Ax^q dx}{\sqrt{2cx - x^2}}$, which may be reduced and integrated as in the preceding article.

186. By the application of the formula for integration by parts, Art. (169), to the expression Xdx, we obtain

$$\int Xdx = Xx - \int xdX \ldots\ldots\ldots\ldots\ldots (1),$$

and then to xdX, &c.,

$$\int xdX = \int \frac{dX}{dx} xdx = \frac{x^2}{2}\frac{dX}{dx} - \int \frac{x^2}{2}\frac{d^2X}{dx} \ldots\ldots (2),$$

$$\int \frac{x^2}{2}\frac{d^2X}{dx} = \int \frac{d^2X}{dx^2}\frac{x^2dx}{2} = \frac{x^3}{2 . 3}\frac{d^2X}{dx^2} - \int \frac{x^3}{2 . 3}\frac{d^3X}{dx^2},$$

..................&c.

Substituting in succession the values above deduced, equation (1) will become

$$\int Xdx = Xx - \frac{dX}{dx}\frac{x^2}{1 . 2} + \frac{d^2X}{dx^2}\frac{x^3}{1 . 2 . 3} - \text{\&c.},$$

a series, expressing the integral of Xdx in terms of X, and its differential coefficients; which has received the name of its distinguished discoverer, John Bernouilli.

187. If in the integral

$$\int Xdx = f(x) = u,$$

we make $x = x + h$, we have

$$(\int Xdx)_{x=x+h} = f(x + h) = u';$$

and, by Taylor's formula,

$$u' - u = \frac{du}{dx}h + \frac{d^2u}{dx^2}\frac{h^2}{1 . 2} + \text{\&c.} \ldots\ldots\ldots (1).$$

But since

$$\int X dx = u, \qquad X dx = du, \qquad \frac{du}{dx} = X,$$

$$\frac{d^2u}{dx^2} = \frac{dX}{dx}, \qquad \frac{d^3u}{dx^3} = \frac{d^2X}{dx^2}, \qquad \&c.$$

These values substituted in (1), give

$$u' - u = Xh + \frac{dX}{dx}\frac{h^2}{1.2} + \frac{d^2X}{dx^2}\frac{h^3}{1.2.3} + \&c.$$

If in this series we make $x = a$, $h = b - a$, and denote by A, A′, A″, &c., what X, $\frac{dX}{dx}$, $\frac{d^2X}{dx^2}$, &c., become under this supposition, it is plain that what u becomes will represent the value of the integral when $x = a$; what u' becomes, its value when $x = a + b - a = b$; then what $u' - u$ becomes, will be the value of the integral between the limits $x = a$, and $x = b$; whence

$$\int_a^b X dx = A(b-a) + \frac{A'}{1.2}(b-a)^2 + \frac{A''}{1.2.3}(b-a)^3 + \&c.,$$

a series from which the approximate value of a definite integral may be obtained. If $b - a$ is so large that the series does not converge, or does not converge rapidly enough, then let it be divided into n equal parts, so that

$$b - a = n\alpha,$$

and take the value, first between the limits a and $a + \alpha$, then between $a + \alpha$ and $a + 2\alpha$, &c., and suppose the results to be

$$\left.\begin{array}{l} B\alpha + B'\frac{\alpha^2}{1.2} + B''\frac{\alpha^3}{1.2.3} + \&c. \\ C\alpha + C'\frac{\alpha^2}{1.2} + C''\frac{\alpha^3}{1.2.3} + \&c. \\ D\alpha + D'\frac{\alpha^2}{1.2} + D''\frac{\alpha^3}{1.2.3} + \&c. \end{array}\right\} \ldots (2),$$

$$\&c.;$$

then, by article (160), we have

$$\int_a^b Xdx = (B + C + D + \&c.)\alpha + (B' + C' + \&c.)\frac{\alpha^2}{1.2} + \&c. \ldots (3),$$

and as α is arbitrary, the separate series (2) [and of course the final series (3)] may be made to converge as rapidly as we please.

INTEGRATION OF DIFFERENTIALS CONTAINING TRANSCENDENTAL QUANTITIES.

188. But few of these differentials admit of exact integrals. We can, however, by the aid of formulas previously deduced, obtain, by series, their approximate integrals.

By the examination of a few expressions, we will endeavor, as far as possible, to indicate to the pupil the general method to be pursued, and then leave to his ingenuity and industry its application to the different cases with which he may meet.

Take first the expression

$$X(lx)^n dx,$$

in which X is an algebraic function of x. If we divide it into the two factors

$$Xdx = dv, \qquad \text{and} \qquad (lx)^n = u;$$

whence

$$\int Xdx = v = X', \qquad du = n(lx)^{n-1}\frac{dx}{x};$$

and then substitute in the formula of Art. (169), we have

$$\int X(lx)^n dx = X'(lx)^n - n\int X'(lx)^{n-1}\frac{dx}{x} \ldots\ldots (1).$$

By this the integral of the primitive expression, when the integral of Xdx can be found, is made to depend upon the integral of another similar one, in which the exponent of (lx) is one less than at first.

If, then, n be entire and positive, after repeated applications of the formula, the exponent of (lx) will become 0, and the expression upon which the integral depends, algebraic.

For a particular case, let

$$X = x^m, \qquad \text{then} \qquad \int x^m dx = \frac{x^{m+1}}{m+1} = X',$$

and this in (1) will give

$$\int x^m(lx)^n dx = \frac{x^{m+1}}{m+1}(lx)^n - \frac{n}{m+1}\int x^m(lx)^{n-1}dx \ldots (2).$$

If in this we substitute for n, in succession,

$$n-1, \qquad n-2, \qquad n-3, \quad \text{\&c.},$$

we have

$$\int x^m (lx)^{n-1} dx = \frac{x^{m+1}}{m+1}(lx)^{n-1} - \frac{n-1}{m+1}\int x^m (lx)^{n-2} dx,$$

$$\int x^m (lx)^{n-2} dx = \frac{x^{m+1}}{m+1}(lx)^{n-2} - \frac{n-2}{m+1}\int x^m (lx)^{n-3} dx,$$

....................&c.

These values in (2), will give a general formula, in which, if n be positive and entire, the last term will be

$$\pm \frac{n(n-1).....2.1}{(m+1)^n}\int x^m (lx)^0 dx = \pm \frac{n(n-1).....1\, x^{m+1}}{(m+1)^{n+1}}.$$

We shall therefore have

$$\int x^m (lx)^n dx = \frac{x^{m+1}}{m+1}\left[(lx)^n - \frac{n(lx)^{n-1}}{m+1} + ... \pm \frac{n(n-1)...1}{(m+1)^n}\right] \,..(3).$$

The sign of the last term will be plus when n is even, and minus when n is odd.

If $m = 1$ and $n = 1$, we have

$$\int xlxdx = \frac{x^2}{2}\left(lx - \frac{1}{2}\right).$$

If $m = 0$ and $n = 1$, we have

$$\int lxdx = x(lx - 1).$$

If $m = -1$, the second member of (3) becomes infinite. In this case the differential becomes

$$(lx)^n \frac{dx}{x}.$$

Making $lx = z$, we have $\frac{dx}{x} = dz$, and

$$\int (lx)^n \frac{dx}{x} = \int z^n dz = \frac{z^{n+1}}{n+1} = \frac{(lx)^{n+1}}{n+1},$$

which is true for all values of n, except when $n = -1$. In this case the expression becomes

$$\frac{dx}{xlx}.$$

Making $lx = z$, we have $\frac{dx}{x} = dz$, and

$$\int \frac{dx}{xlx} = \int \frac{dz}{z} = lz = l\,(lx).$$

189. Take now the expression

$$Xa^x dx,$$

in which X is an algebraic function of x. If we divide it into the two factors X and $a^x dx$, and recollect that

$$a^x ladx = da^x \ldots\ldots\ldots \text{Art. (39)},$$

whence

$$a^x dx = \frac{da^x}{la}, \qquad \text{and} \qquad \int a^x dx = \frac{a^x}{la};$$

we shall have, from the formula for integration by parts,

$$\int Xa^x dx = \frac{Xa^x}{la} - \int \frac{a^x}{la} dX \ldots\ldots\ldots (1).$$

If we take the successive differentials of X, and place

$$dX = X'dx, \qquad dX' = X''dx, \qquad dX'' = X'''dx, \quad \&c.,$$

we obtain

$$\int \frac{a^x dX}{la} = \frac{X'a^x}{(la)^2} - \int \frac{a^x}{(la)^2} dX',$$

$$\int \frac{a^x dX'}{(la)^2} = \frac{X''a^x}{(la)^3} - \int \frac{a^x}{(la)^3} dX'',$$

&c.

These values, in equation (1), give

$$\int Xa^x dx = a^x \left(\frac{X}{la} - \frac{X'}{(la)^2} \cdots\cdots \frac{\pm X^{n'}}{(la)^{n+1}} \right) \mp \int \frac{a^x dX^{n'}}{(la)^{n+1}} \cdots\cdots (2).$$

If the function X is of such a nature that one of its differential coefficients X', X'', &c., is constant, the differential of this will be 0, and the corresponding term

$$\mp \int \frac{a^x dX^{n'}}{(la)^{n+1}} = 0.$$

The integral will then be exact.

The expression $$x^n a^x dx,$$

admits of an exact integral when n is entire and positive.

If n be fractional or negative, we write for a^x its development, Art. (40), and then integrate as in Art. (185).

190. By article (43), we have

$$d \sin nx = ndx \cos nx, \qquad d \cos nx = -ndx \sin nx;$$

hence

$$\int dx \sin nx = -\frac{\cos nx}{n}, \qquad \int dx \cos nx = \frac{\sin nx}{n}.$$

In the expression

$$dx \sin^2 x,$$

we can place for $\sin^2 x$, its value, $\frac{1}{2} - \frac{\cos 2x}{2}$, and then have

$$\int dx \sin^2 x = \int \frac{dx}{2} - \int \frac{\cos 2x dx}{2} = \frac{x}{2} - \frac{1}{4} \sin 2x;$$

and, in general, the integral of similar expressions, containing any power of either the sine or cosine of x, can be obtained by first substituting the value of the power in terms of the functions of the double, triple, &c., arc, as determined in Trigonometry.

The expressions

$$dx \sin^m x, \qquad dx \cos^m x,$$

when m is entire, may also be integrated as follows: Make

$$\sin x = z, \qquad \text{then} \qquad x = \sin^{-1} z, \quad dx = \frac{dz}{(1 - z^2)^{\frac{1}{2}}};$$

whence

$$\int dx \sin^m x = \int \frac{z^m dz}{(1 - z^2)^{\frac{1}{2}}}.$$

This expression, by repeated applications of formula **A** or **C**, may be made to depend upon

$$\int \frac{dz}{(1 - z^2)^{\frac{1}{2}}}, \qquad \text{or} \qquad \int \frac{zdz}{(1 - z^2)^{\frac{1}{2}}}.$$

In the expression

$$dx \tan^m x,$$

place
$$\tan x = z,$$

hen

$$dx = \frac{dz}{1+z^2}, \qquad \int dx \tan^m x = \int \frac{z^m dz}{1+z^2},$$

which is a rational fraction.

Examples.

Integrate

1. $du = dx \sin^3 x.$ 2. $du = \frac{dx}{\cos^3 x}.$

3. $du = \frac{dx}{\sin x}.$ 4. $du = dx \tan^2 x.$

191. In the general expression

$$dx \sin^m x \cos^n x,$$

we may place

$$\sin x = z, \quad \text{then} \quad \cos x = (1 - z^2)^{\frac{1}{2}}, \quad dx = \frac{dz}{(1-z^2)^{\frac{1}{2}}},$$

and finally,

$$\int dx \sin^m x \cos^n x = \int z^m dz (1 - z^2)^{\frac{n-1}{2}},$$

which may be reduced by formulas **A**, **B**, **C**, and **D**, and in some cases integrated, as in the example

$$du = dx \sin^4 x \cos^2 x; \quad \text{whence} \quad u = \int z^4 dz (1 - z^2)^{\frac{1}{2}}.$$

192. Take finally the expression

$$\mathrm{X} dx \sin^{-1} x.$$

Place $\quad \mathrm{X}dx = dv, \quad$ and $\quad \sin^{-1}x = u, \quad$ then

$$\int \mathrm{X}dx = v = \mathrm{X}', \quad \text{and} \quad du = \frac{dx}{(1 - x^2)^{\frac{1}{2}}}.$$

Substituting in the formula of Art. (169), we have

$$\int \mathrm{X}dx \sin^{-1}x = \mathrm{X}' \sin^{-1}x - \int \frac{\mathrm{X}'dx}{(1 - x^2)^{\frac{1}{2}}}.$$

Thus the integral of the primitive expression is made to depend upon the integral of the algebraic expression $\dfrac{\mathrm{X}'dx}{(1 - x^2)^{\frac{1}{2}}}$.

Let $$\mathrm{X} = x^n,$$

then

$$\int \mathrm{X}dx = \int x^n dx = \frac{x^{n+1}}{n + 1} = \mathrm{X}',$$

and we have

$$\int x^n dx \sin^{-1}x = \frac{x^{n+1}}{n + 1} \sin^{-1}x - \frac{1}{n + 1} \int \frac{x^{n+1}dx}{(1 - x^2)^{\frac{1}{2}}}.$$

By the application of formula **A** or **C**, when n is entire, the last term may be reduced, and then integrated; except when $n = -1$, in which case the expression becomes

$$\frac{dx}{x} \sin^{-1}x,$$

which can only be integrated by series.

In the same way, like expressions may be found for

$$\int Xdx \cos^{-1}x, \qquad \int Xdx \tan^{-1}x, \text{ \&c.}$$

Integration of Differentials of the Higher Orders.

193. By an application of the rules previously demonstrated, we may readily obtain the primitive function, from which differentials, of a higher order than the first, containing a single variable, may have been derived.

Let there be the differential

$$d^n u = Xdx^n,$$

X being any function of x.

Dividing by dx^{n-1}, we have

$$\frac{d^n u}{dx^{n-1}} = Xdx,$$

and since dx^{n-1} is a constant, this may be written, Art. (26),

$$d\left(\frac{d^{n-1}u}{dx^{n-1}}\right) = Xdx.$$

Integrating both members, we have

$$\frac{d^{n-1}u}{dx^{n-1}} = \int X dx = X' + C.$$

After multiplying both members of this equation by dx, it may be written

$$d\left(\frac{d^{n-2}u}{dx^{n-2}}\right) = X'dx + Cdx;$$

and integrating as before,

$$\frac{d^{n-2}u}{dx^{n-2}} = X'' + Cx + C';$$

which by another transformation and integration, may be reduced one degree lower, and finally after n integrations, we shall obtain

$$u = F(x) + \frac{Cx^{n-1}}{1.2..(n-1)} + \frac{C'x^{n-2}}{1.2..(n-2)} + \ldots\ldots\ldots C^{(n-1)'}.$$

The above operation may be indicated thus,

$$u = \int^n X dx^n;$$

the symbol $\int^n$ indicating that n successive integrations are required.

At each integration an arbitrary constant is introduced. The complete integral may therefore be required to fulfil n arbitrary conditions.

Examples.

1. Let $$d^2u = ax^2dx^2.$$

The required operation is indicated thus,

$$u = \int^2 ax^2 dx^2,$$

and may be read, *the double integral of* ax^2dx^2.

Let the expression, after dividing by dx, be written

$$\frac{d^2u}{dx} = d\left(\frac{du}{dx}\right) = ax^2dx;$$

whence, by integration,

$$\frac{du}{dx} = \frac{ax^3}{3} + \text{C}, \qquad du = \frac{ax^3}{3}dx + \text{C}dx.$$

Integrating again, we obtain

$$u = \frac{ax^4}{12} + \text{C}x + \text{C}'.$$

2. If

$$d^3u = bdx^3, \qquad u = \int^3 bdx^3,$$

which is called *a triple integral.* We may write

$$\frac{d^3u}{dx^2} = d\left(\frac{d^2u}{dx^2}\right) = bdx;$$

whence

$$\frac{d^2u}{dx^2} = bx + \text{C};$$

and finally, as in the last example,

$$u = \int^3 bdx^3 = \frac{bx^3}{6} + \frac{\text{C}x^2}{2} + \text{C}'x + \text{C}''.$$

3. $d^4u = -\dfrac{6dx^4}{x^4}$. 4. $d^3u = \sqrt{x}\,dx^3$.

Integration of Partial Differentials.

194. Hitherto, we have explained the mode of integrating only the differentials of functions of a single variable. It yet remains to extend our rules to the integration of those which contain more than one variable.

These differentials are either *partial* or *total*, Art. (52). When partial, they belong to one of *two classes:*

I. Those obtained from the primitive function by differentiating with reference to one variable only.

II. Those obtained by differentiating first with reference to one variable, and then with reference to another, &c., Art. (48).

195. The differentials of the first class may be expressed generally thus,

$$d^n u = f(x, y, z, \&c.)dx^n; \quad d^n u = f'(x, y, z, \&c.)dy^n; \ \&c.,$$

in which u is a function of x, y, z, &c., and may evidently be obtained by successive integrations, precisely as in article (193); all the variables, except the one with reference to which the differentiation was made, being regarded as constant, and care being taken to add, instead of constants, arbitrary functions of those variables which are regarded as constant during the integration.

Examples.

1. Let $$d^2u = bx^2ydx^2,$$

which, after dividing by dx, may be written

$$d\left(\frac{du}{dx}\right) = bx^2ydx;$$

whence

$$\frac{du}{dx} = \int bx^2ydx = \frac{bx^3y}{3} + \text{Y},$$

$$du = \frac{bx^3y}{3}dx + \text{Y}dx,$$

and

$$u = \int^2 bx^2ydx^2 = \frac{bx^4y}{12} + \text{Y}x + \text{Y}',$$

in which Y and Y′ are arbitrary functions of y.

2. Let $$d^3u = cx^2y^2z^2dy^3.$$

196. The differentials of the second class may be written generally thus,

$$d^{m+n+\cdots}u = f(x, y, z, \&c.)\, dx^m dy^n dz^p \ldots\ldots\ldots,$$

and the mode of integrating is plainly to integrate first, m times with reference to x, then n times with reference to y, and so on until all the required integrations are made.

To illustrate, let

1. $$d^2u = 2x^2y\,dx\,dy,$$

which may be written

$$d\left(\frac{du}{dx}\right) = 2x^2y\,dy,$$

whence, by integration with reference to y,

$$\frac{du}{dx} = x^2y^2 + X, \qquad \text{or} \qquad du = x^2y^2dx + Xdx,$$

and

$$u = \int^2 2x^2y\,dx\,dy = \frac{x^3y^2}{3} + \int Xdx + Y;$$

there being no necessity of indicating with reference to which variable the integration is first to be made, Art. (49).

2. Let $$d^3u = ax^2ydy^2dx.$$

This may be written

$$\frac{d^3u}{dy^2} = ax^2ydx, \qquad \text{or} \qquad d\left(\frac{d^2u}{dy^2}\right) = ax^2ydx.$$

Integrating with reference to x,

$$\frac{d^2u}{dy^2} = \frac{ax^3y}{3} + Y,$$

which may now be integrated as in the preceding article.

3. $d^3u = axz^2dxdydz.$ 4. $d^4u = (x + y)^2dx^2dy^2.$

INTEGRATION OF TOTAL DIFFERENTIALS OF THE FIRST ORDER.

197. If $$u = f(x, y),$$

we have found, Art. (52),

$$du = \frac{du}{dx}dx + \frac{du}{dy}dy,$$

in which $\frac{du}{dx}dx$ and $\frac{du}{dy}dy$ are the partial differentials of $f(x, y)$; and also, Art. (49),

$$\frac{d^2u}{dxdy} = \frac{d^2u}{dydx}, \quad \text{or} \quad \frac{d\left(\frac{du}{dx}\right)}{dy} = \frac{d\left(\frac{du}{dy}\right)}{dx} \ldots\ldots(1).$$

If, then, an expression of the form

$$\mathrm{P}dx + \mathrm{Q}dy \ldots\ldots\ldots\ldots\ldots\ldots(2),$$

be the total differential of a function of x and y; Pdx and Qdy must be the two partial differentials of the function, and by the integration of either, we shall obtain the function itself.

To ascertain, in any given expression of the above form, whether Pdx and Qdy are such partial differentials, we have simply to see if the condition (1), or

$$\frac{d\mathrm{P}}{dy} = \frac{d\mathrm{Q}}{dx},$$

is fulfilled. If so, the given expression is the differential of a function of x and y, and we have

$$u = \int \mathrm{P}dx + \mathrm{Y} \ldots\ldots\ldots\ldots\ldots(3),$$

Y being a function of y, which is to be determined so as to satisfy the condition $\frac{du}{dy} = \mathrm{Q}$.

Since the differential of every term of u which contains x, when taken with respect to x, must contain dx, the integral of Pdx will give all that part of u which contains x. The differential of those terms which contain y and do not contain x, will be found only in the expression Qdy. If, then, we integrate those terms of Qdy which do not contain x, we shall have that part of u which con tains y and not x. This will be the required expression for Y which added, with an arbitrary constant, to $\int P dx$, will give the complete integral.

If all the terms of the given differential contain x or dx, Y will be 0, and we complete the integral by the addition of an arbitrary constant to the integral of Pdx.

Examples.

1. Let

$$du = (2axy - 3bx^2y)dx + (ax^2 - bx^3)dy,$$

which, compared with equation (2), gives

$$P = 2axy - 3bx^2y, \qquad Q = ax^2 - bx^3,$$

$$\frac{dP}{dy} = 2ax - 3bx^2 = \frac{dQ}{dx}.$$

This condition being fulfilled, we then have, since all the terms of du contain x or dx,

$$u = \int(2axy - 3bx^2y)dx = ax^2y - byx^3 + C.$$

2. If

$$du = \frac{dx}{y} + \left(2y - \frac{x}{y^2}\right)dy,$$

$$u = \int\frac{dx}{y} = \frac{x}{y} + Y.$$

The term of Qdy which does not contain x, is

$$2y\,dy,$$

the integral of which is y^2; hence

$$Y = y^2,$$

and the above expression for u becomes

$$u = \frac{x}{y} + y^2 + C.$$

3. If $$du = \frac{ydx - xdy}{x^2 + y^2}, \qquad u = \tan^{-1}\frac{x}{y} + C.$$

4. Let $$du = (6xy - y^2)\,dx + (3x^2 - 2xy)\,dy.$$

198. The method of obtaining the integral of a differential, containing several variables, is readily deduced from what precedes. Let

$$du = Pdx + Qdy + Rdz = df(x, y, z)\ldots.(1).$$

If for a moment we regard z as a constant, and then, in succession, y and x, it is plain that we shall have the three expressions

$$Pdx + Qdy, \qquad Pdx + Rdz, \qquad Qdy + Rdz\ldots(2),$$

which, taken separately, are differentials of functions of two variables, if the primitive expression is a differential of a function of three, and the reverse.

But the conditions that these be each an exact differential, are

$$\frac{dP}{dy} = \frac{dQ}{dx}, \qquad \frac{dP}{dz} = \frac{dR}{dx}, \qquad \frac{dQ}{dz} = \frac{dR}{dy}\ldots.(3);$$

hence, if we have given an expression of the form

$$Pdx + Qdy + Rdz,$$

and the conditions (3) are fulfilled, it will be the differential of a function of three variables, and we can obtain the function by integrating either of the expressions (2), as in the preceding article, taking care to add to the integral a function of that variable which is regarded as constant. Thus, denoting the integral of $Pdx + Qdy$ by v, we have

$$\int(Pdx + Qdy + Rdz) = v + Z \ldots\ldots (4),$$

Z being independent of x and y, and a function of z alone; and may be determined by taking the integral of those terms of Rdz which contain neither x nor y.

199. If a function of two variables, composed of entire terms, is homogeneous with reference to the variables, its differential will also be homogeneous; and such a relation will exist between the function and its partial differential coefficients, as will enable us at once to obtain the function when the differential is given.

To explain this relation, let

$$u = f(x, y),$$

and m denote the sum of the exponents of x and y in each term. For x and y, substitute tx and ty respectively; the primitive function then becomes $t^m u$.

In this expression, for t put $(1 + s)$; then

$$t^m u = (1 + s)^m u.$$

Under these suppositions, x and y, in the primitive function, have become, respectively, $x + sx$, and $y + sy$.

Developing this new state of the primitive function, as in article (48), we have

$$+ \left(\frac{du}{dx} sx + \frac{du}{dy} sy\right) + \frac{1}{2}\left(\frac{d^2u}{dx^2} s^2x^2 + 2\frac{d^2u}{dxdy} s^2xy\ldots\right) + \&c.$$

$$= (1 + s)^m u = u + mus + \frac{m(m - 1)us^2}{1.2} + \&c.$$

Equating the coefficients of the first powers of the indeterminate s, we have

$$\frac{du}{dx}x + \frac{du}{dy}y = mu\ldots\ldots.(1).$$

Hence, in the differential

$$du = \mathrm{P}dx + \mathrm{Q}dy,$$

if P and Q are homogeneous of the $(m - 1)$th degree, we shall have, by comparison with equation (1),

$$\mathrm{P}x + \mathrm{Q}y = mu, \qquad u = \frac{\mathrm{P}x + \mathrm{Q}y}{m}.$$

For example, let

$$du = 4xy^2dx + ay^3dx + 4x^2ydy + 3axy^2dy,$$

in which $m - 1 = 3, \quad m = 4,$

$$4xy^2 + ay^3 = \mathrm{P}, \qquad 4x^2y + 3axy^2 = \mathrm{Q};$$

whence

$$u = \frac{\mathrm{P}x + \mathrm{Q}y}{4} = 2x^2y^2 + axy^3.$$

200. If we denote $\int \mathrm{P}dx$ by v, we have, by passing to the differential coefficient,

$$\frac{dv}{dx} = \mathrm{P}.$$

Differentiating this with reference to the variable y, we find

$$\frac{d\left(\dfrac{dv}{dx}\right)}{dy} = \frac{dP}{dy} = \frac{d\left(\dfrac{dv}{dy}\right)}{dx} \ldots\ldots \text{Art. (197)};$$

whence

$$\frac{d\left(\dfrac{dv}{dy}\right)}{dx}dx = \frac{dP}{dy}dx.$$

Integrating with reference to the variable x, we have

$$\frac{dv}{dy} = \int \frac{dP}{dy}dx,$$

or, since $(dP)\,dx = d\,(Pdx)$,

$$\frac{d\int Pdx}{dy} = \int \frac{d\,(Pdx)}{dy}.$$

By which we see, *that we may differentiate with reference to another variable, the indicated integral of a partial differential, by simply differentiating the quantity under the sign.*

Integration of Differential Equations.

201. These equations, when of the first order, and when derived from equations containing but two variables, will appear as particular cases of the general form

$$Pdx + Qdy = 0,$$

and may of course be integrated as in article (197), when

$$\frac{dP}{dy} = \frac{dQ}{dx},$$

and give

$$\int Pdx + Y = C.$$

In practice, however, it will in general be found that, in consequence of the disappearance of a factor common to both terms of the differential equation, or when the differential equation has been obtained by the elimination of a constant from the primitive and its immediate differential equation, Art. (58), this condition is not fulfilled; hence other means of obtaining the integral must be sought for.

In the first place, it is evident that, if by any transformation the equation can be placed under the form

$$Xdx + Ydy = 0,$$

X being a function of x, and Y of y, the integral can be found by taking the sum of the integrals of the two terms; thus,

$$\int Xdx + \int Ydy = C.$$

202. Among the most simple forms with which we meet, are

I. $$Ydx + Xdy = 0.$$

II. $$XYdx + X'Y'dy = 0.$$

The variables may be separated, in I., by dividing by YX; and in II., by dividing by YX'. The results,

$$\frac{dx}{X} + \frac{dy}{Y} = 0,$$

and

$$\frac{X}{X'}dx + \frac{Y'}{Y}dy = 0,$$

are under the proposed form. In general, if the value of $\frac{dy}{dx}$, deduced from the equation, be under the form

$$\frac{dy}{dx} = \mathrm{XY},$$

we have

$$\frac{dy}{\mathrm{Y}} = \mathrm{X}dx, \qquad \text{and} \qquad \int\frac{dy}{\mathrm{Y}} = \int \mathrm{X}dx.$$

Examples.

1. Let $$ydx - xdy = 0.$$

Dividing by yx, we have

$$\frac{dx}{x} - \frac{dy}{y} = 0, \qquad lx - ly = \mathrm{C};$$

or, making $\mathrm{C} = l\mathrm{C}'$, we have

$$l\frac{x}{y} = l\mathrm{C}', \qquad \frac{x}{y} = \mathrm{C}', \qquad x = \mathrm{C}'y.$$

2. Let $$xy^2dx + dy = 0.$$

Dividing by y^2,

$$xdx + \frac{dy}{y^2} = 0;$$

integrating, and reducing,

$$x^2y - 2 = 2\mathrm{C}y.$$

3. Let $(1 - x)^2 ydx - (1 + y)x^2 dy = 0;$

whence

$$\frac{(1 - x)^2}{x^2}dx - \frac{1 + y}{y}dy = 0,$$

and

$$-\frac{1}{x} - 2lx + x - ly - y = C.$$

4. Let $(1 + x^2)dy - \sqrt{y}\,dx = 0.$

5. Let $x^2 ydx - (3y + 1)\sqrt{x^3}\,dy = 0.$

203. III. In all cases where the equation is homogeneous with reference to the variables, they can be separated, and the equation placed under the proposed form.

Suppose the general form of the given differential to be

$$Ax^n y^m dx + Bx^h y^k dy = 0,$$

in which $n + m = h + k = g.$

Make $y = zx$, and substitute; we thus obtain

$$Ax^g z^m dx + Bx^g z^k dy = 0;$$

dividing by x^g, and putting for dy its value, $zdx + xdz$, we have

$$Az^m dx + Bz^k(zdx + xdz) = 0;$$

dividing by $(Az^m + Bz^{k+1})x$, we have

$$\frac{dx}{x} + \frac{Bz^k dz}{Az^m + Bz^{k+1}} = 0,$$

which is under the proposed form.

Examples.

1. Let $$x^2dy - y^2dx - xydx = 0.$$

Make $y = zx$, then $dy = zdx + xdz$.

Substituting in the given equation, we have

$$x^2zdx + x^3dz - z^2x^2dx - x^2zdx = 0;$$

reducing and integrating,

$$xdz - z^2dx = 0, \qquad -\frac{1}{z} - lx = \text{C}.$$

Putting for z its value, we have finally

$$lx = -\left(\text{C} + \frac{x}{y}\right).$$

2. If $\dfrac{x^2 + yx}{x - y}dy = ydx, \qquad lx = \dfrac{x}{2y} - l\sqrt{\dfrac{y}{x}} + \text{C}.$

3. Let $$xdy - ydx = dx\sqrt{x^2 + y^2}.$$

204. IV. The equation

$$(a + bx + cy)dx + (a' + b'x + c'y)dy = 0,$$

may be so transformed, that the variables can be separated and the integral found. For this purpose let us make

$$x = t + \delta, \qquad y = u + \delta';$$

whence

$$dx = dt, \qquad dy = du.$$

These values in the primitive equation, give

$$(a + b\delta + c\delta' + bt + cu)dt + (a' + b'\delta + c'\delta' + b't + c'u)du = 0.$$

By placing

$$a + b\delta + c\delta' = 0, \qquad a' + b'\delta + c'\delta' = 0,$$

we can determine proper values for the arbitrary quantities δ and δ', and our equation reduces to

$$(bt + cu)dt + (b't + c'u)du = 0;$$

which being homogeneous with reference to t and u may be treated as in the preceding article.

This transformation is always possible, save when the values of δ and δ' become infinite, which will be the case only when

$$bc' - cb' = 0;$$

whence

$$c' = \frac{cb'}{b}; \qquad b'x + c'y = \frac{b'}{b}(bx + cy).$$

The primitive equation thus becomes

$$adx + a'dy + (bx + cy)\left(dx + \frac{b'}{b}dy\right) = 0,$$

in which the variables may be separated by making

$$bx + cy = z.$$

Substituting this, and the resulting value of dy, the equation reduces to

$$dx + \frac{(a'b + b'z)\,dz}{abc - a'b^2 + (bc - bb')\,z} = 0.$$

If
$$bc - bb' = 0,$$

we have at once the integral

$$x + \frac{2a'bz + b'z^2}{2(abc - a'b^2)} = \mathrm{C},$$

in which the value of z is to be substituted.

205. V. In the equation

$$dy + \mathrm{P}y\,dx = \mathrm{Q}dx \ldots\ldots (1),^*$$

P and Q being functions of x, the variables may be readily separated by making

$$y = z\mathrm{X} \ldots\ldots\ldots\ldots (2),$$

X being a function of x, for which a proper value is to be determined. By differentiating equation (2), we have

$$dy = zd\mathrm{X} + \mathrm{X}dz,$$

and by substitution in (1),

$$zd\mathrm{X} + \mathrm{X}(dz + \mathrm{P}z\,dx) = \mathrm{Q}dx \ldots (3).$$

Suppose X to have such a value that

$$zd\mathrm{X} = \mathrm{Q}dx \ldots\ldots\ldots\ldots (4);$$

---

* Equations of this kind, being of the first degree with reference to y and dy, are sometimes improperly called *linear equations*.

equation (3) then becomes

$$X(dz + Pzdx) = 0;$$

whence

$$\frac{dz}{z} = -Pdx, \qquad lz = -\int Pdx;$$

or, taking the numbers,

$$z = e^{-\int Pdx}.$$

From equation (4), we have

$$dX = \frac{Qdx}{z} = Qe^{\int Pdx}dx;$$

whence

$$X = \int Qe^{\int Pdx}dx.$$

These values of z and X, in equation (2), give

$$y = e^{-\int Pdx}\int Qe^{\int Pdx}dx.$$

Example.

Let $$dy + ydx = e^{-x}dx,$$

then $$P = 1, \qquad Q = e^{-x}, \qquad \int Pdx = x;$$

hence, by substitution in the above value of y,

$$y = e^{-x}\int e^{-x}.e^{x}dx = e^{-x}(x + C).$$

206. If we divide the form

$$dy + Pydx = Qy^m dx \ldots\ldots(1),$$

by y^m, we have

$$\frac{dy}{y^m} + Py^{-m+1}dx = Qdx\ldots\ldots(2),$$

In this make

$$y^{-m+1} = z, \qquad \text{whence} \qquad -(m-1)y^{-m}dy = dz,$$

and

$$dy = -\frac{y^m dz}{m-1}.$$

Substituting these values in (2), and reducing, we obtain

$$dz - (m-1)Pzdx = -(m-1)Qdx,$$

the same form as equation (1) of the preceding article. Integrating this, and substituting the value of z, we shall have the primitive equation, from which equation (1) may be derived.

207. Equations of the form

$$ay^m x^n dx + bx^p dx + cx^q dy = 0,$$

may sometimes be rendered homogeneous by making

$$y = z^k,$$

k being a constant to be determined. From this, we have

$$dy = kz^{k-1}dz, \qquad y^m = z^{km}.$$

These values in the primitive equation give

$$az^{km}x^n dx + bx^p dx + ckx^q z^{k-1} dz = 0,$$

which will be homogeneous, if

$$km + n = p = q + k - 1;$$

that is, when

$$\frac{p - n}{m} = p + 1 - q = k.$$

Of the Factors by which certain Differential Equations are rendered Integrable.

208. It has been remarked, article (201), that differential equations sometimes fail to fulfil the condition of integrability, in consequence of the disappearance of a common factor. Whenever this factor can be discovered, by trial or otherwise, the integral can at once be found, as in article (197).

Let $$\text{P}dx + \text{Q}dy = 0,$$

be a differential equation in which the condition is not fulfilled, and suppose that

$$z = f(x, y),$$

is the factor by the disappearance of which the given equation has resulted. The immediate differential equation will then be

$$\text{P}zdx + \text{Q}zdy = 0,$$

from which we have the condition, Art. (197),

$$\frac{d\text{P}z}{dy} = \frac{d\text{Q}z}{dx};$$

or, performing the differentiation,

$$\frac{zdP}{dy} + \frac{Pdz}{dy} = \frac{zdQ}{dx} + \frac{Qdz}{dx},$$

or

$$\left(P\frac{dz}{dy} - Q\frac{dz}{dx}\right) + \left(\frac{dP}{dy} - \frac{dQ}{dx}\right)z = 0 \ldots\ldots (1).$$

This equation expresses a relation between z, x, and y, but its solution in the general case is so difficult, that nothing will be gained by attempting it.

209. If it be possible to find one factor which will render the differential equation integrable, an infinite number of others will at once result. For, suppose an expression for z has been found; then

$$zPdx + zQdy = du$$

is a differential which is integrable. If we multiply both members by *any arbitrary function of* u, denoted by U, we have

$$UzPdx + UzQdy = Udu.$$

Udu, containing u alone, is a differential; hence the first member is also a differential of some function of x and y, which admits of integration; and zU, or

$$zF\int(zPdx + zQdy),$$

is a factor which will render the given differential equation integrable.

210. In the particular case where z is a function of x only, its value can be determined, as we shall then have

$$\frac{dz}{dy} = 0,$$

and equation (1), of Art. (208), will reduce to

$$-\frac{Qdz}{dx} + \left(\frac{dP}{dy} - \frac{dQ}{dx}\right)z = 0,$$

or

$$\frac{dz}{z} = \frac{1}{Q}\left(\frac{dP}{dy} - \frac{dQ}{dx}\right)dx.$$

But by hypothesis z is a function of x, therefore

$$\frac{1}{Q}\left(\frac{dP}{dy} - \frac{dQ}{dx}\right) = f(x) = X;$$

then

$$\int\frac{dz}{z} = \int Xdx;$$

whence

$$lz = \int Xdx, \qquad z = e^{\int Xdx}.$$

Let this be illustrated by the example

$$dx + 2xydy + 2y^2dx = 0,$$

in which

$$P = 1 + 2y^2, \qquad Q = 2xy;$$

whence

$$\frac{1}{Q}\left(\frac{dP}{dy} - \frac{dQ}{dx}\right) = \frac{1}{x} = X,$$

and

$$z = e^{\int Xdx} = e^{\int\frac{1}{x}dx} = e^{lx} = x,$$

x being the common factor, the immediate differential equation must be

$$xdx + 2x^2ydy + 2xy^2dx = 0,$$

which can be integrated as in article (197).

In a similar way, if $z = f'(y)$, its value may be determined.

To ascertain whether a proper expression for z has been thus obtained, we multiply by it, and then apply the test as given in Art. (197).

211. If the given differential equation be homogeneous with respect to the variables, a proper expression for z may be found. Let

$$Pdx + Qdy = 0$$

be homogeneous and of the $m - 1^{th}$ degree, and suppose z is of the nth degree; then

$$zPdx + zQdy = 0,$$

will be of the $m - 1 + n^{th}$ degree. Hence, by Art. (199), we have

$$\int (zPdx + zQdy) = \frac{zPx + zQy}{m + n} = C;$$

whence

$$z = \frac{C(m + n)}{Px + Qy} = \frac{1}{Px + Qy},$$

since, C being arbitrary, we may make

$$C(m + n) = 1.$$

212. If the given differential equation can be divided into two parts, and a separate factor can be found which will render each part integrable, a third factor may sometimes be deduced from these, which will render the given equation integrable. Thus, let

$$Pdx + Qdy = 0$$

be divided into the two parts

$$(P'dx + Q'dy) + (P''dx + Q''dy) = 0,$$

and suppose z' to be a factor which will render the first part integrable, and z'' a like factor for the second part; then

$$z'P'dx + z'Q'dy = du', \qquad z''P''dx + z''Q''dy = du'';$$

from which u' and u'' may be obtained as in Art. (197). Then, from Art. (209), $z'U'$ and $z''U''$ are also factors which will render the respective parts integrable, U' and U'' being arbitrary functions of u' and u''. If, therefore, we can assign, by trial or otherwise, such values to U' and U'' as to make

$$z'U' = z''U'',$$

the expression resulting will be a factor which will render the two parts integrable, and of course the given equation.

Integration of Differential Equations containing the Higher Powers of dy.

213. Differential equations of the first order, containing the higher powers of dy, may arise, as in the third example of article (58), from the elimination of the higher powers of a constant. Such equations, after division by dx^n, may be put under the form

$$\left(\frac{dy}{dx}\right)^n + M\left(\frac{dy}{dx}\right)^{n-1} \dots\dots\dots + U = 0 \dots\dots (1).$$

The determination of the primitive equation will then depend upon the solution of equation (1), or upon the division of the first member into its factors of the first degree. There are n such factors, and it is plain that each, when placed equal to zero and integrated, will give an equation between y and x which may be regarded as a primitive equation.

If, then, the values of $\frac{dy}{dx}$ be denoted by V, V′, V″, &c., equation (1) may be written

$$\left(\frac{dy}{dx} - \text{V}\right)\left(\frac{dy}{dx} - \text{V}'\right)\left(\frac{dy}{dx} - \text{V}''\right)\text{\&c.} = 0,$$

which may be satisfied by placing

$$\frac{dy}{dx} - \text{V} = 0, \qquad \frac{dy}{dx} - \text{V}' = 0, \quad \text{\&c.}\ldots\ldots(2);$$

and if the corresponding primitive equations be denoted by $\text{P} = 0$, $\text{P}' = 0$, $\text{P}'' = 0$, &c., respectively, we shall have

$$\text{PP}'\text{P}''\text{\&c.} = 0 \ldots\ldots\ldots\ldots\ldots(3),$$

for the most general primitive equation, particular cases of which may be obtained by placing $\text{P} = 0$, $\text{P}' = 0$, or the product of any of these factors taken two and two, or three and three, &c.

It would appear, since a constant is to be added in the integration of each of the equations (2), that (3) ought to contain n arbitrary constants; but equation (1) can only be deduced from its primitive equation by the elimination of the nth power of a constant: [Or by raising $\left(\frac{dy}{dx} - \text{V}\right)$ to the nth power, in which case the primitive equation must be $y = \int \text{V}dx$]. It is plain, then, that the constants added ought to be equal, or that the same should be added in each integration.

The n differential equations of the first degree which are factors of (1), are readily accounted for, by supposing the primitive equation to be solved with reference to C, which will have n values, each of which, differentiated, will give one of the equations referred to.

As there will be difficulty in the solution of equation (1), when the degree is higher than the second, it will be well to discuss some particular cases which admit of integration by other means.

214. I. If the proposed equation does not contain y, and it be easier to solve it with reference to x than with reference to $\frac{dy}{dx}$, which we will denote by p, we can then obtain

$$x = \varphi(p) \ldots\ldots\ldots\ldots (1).$$

But
$$dy = pdx,$$

and by parts, article (169),

$$y = px - \int xdp = px - \int \varphi(p)\,dp + C;$$

whence, if $\varphi(p)\,dp$ can be integrated, p may be eliminated by the aid of equation (1), and the primitive equation between x, y, and C, deduced.

II. If the proposed equation does not contain x, and may be solved with reference to y, we shall have

$$y = f(p) \ldots\ldots\ldots\ldots (3),$$

$$dy = df(p), \qquad \text{or} \qquad pdx = df(p);$$

whence

$$dx = \frac{df(p)}{p}, \qquad x = \int \frac{df(p)}{p} + C.$$

Combining this with equation (3), and eliminating p, a primitive equation will result between x, y, and C.

III. If the proposed equation is homogeneous with respect to the variables of the nth degree, we may make

$$y = tx \ldots\ldots\ldots\ldots (4),$$

and then divide by x^n, and, if possible, solve the equation with respect to t, and have

$$t = f(p) \ldots\ldots\ldots\ldots (5).$$

Differentiating (4), we have

$$dy = xdt + tdx, \qquad \text{or} \qquad pdx = xdt + tdx,$$

$$\frac{dx}{x} = \frac{dt}{p - t} = \frac{df(p)}{p - f(p)},$$

the integral of which is

$$lx = \varphi(p).$$

By combining this with (4) and (5), a primitive equation between y and x may be obtained.

IV. When both variables enter, but y enters only to the first power, we may take its value in terms of p and x, differentiate it, and thus obtain

$$dy = \text{R}dx + \text{S}dp;$$

or, since $dy = pdx$,

$$(\text{R} - p)\,dx + \text{S}dp = 0.$$

If this can be integrated, the result may be combined with the proposed equation, p eliminated, and a primitive equation between y and x determined.

Suppose the deduced value of y to be

$$y = px + \mathrm{P} \dots\dots\dots\dots (6),$$

in which $\mathrm{P} = f(p)$. By differentiation, we obtain

$$dy = pdx + xdp + \frac{d\mathrm{P}}{dp}dp;$$

or

$$\left(x + \frac{d\mathrm{P}}{dp}\right)dp = 0,$$

which may be satisfied by making

$$x + \frac{d\mathrm{P}}{dp} = 0 \dots\dots (7), \qquad \text{or} \qquad dp = 0 \dots\dots (8).$$

Equation (8) gives $\qquad p = \mathrm{C};$

whence, by substitution in (6),

$$y = \mathrm{C}x + \mathrm{C}',$$

C' being what P becomes when $p = \mathrm{C}$.

Equation (7) expresses a relation between x and p, and if it be combined with (6), and p be eliminated, an equation between x and y will result, which will contain no arbitrary constant.

Let there be for a particular example

$$ydx - xdy = n\sqrt{dx^2 + dy^2};$$

whence

$$y = px + n\sqrt{1 + p^2} \dots\dots\dots (9),$$

$$dy = pdx + xdp + \frac{npdp}{\sqrt{1 + p^2}},$$

$$dp\left(x + \frac{np}{\sqrt{1 + p^2}}\right) = 0;$$

whence

$$x + \frac{np}{\sqrt{1 + p^2}} = 0, \qquad dp = 0, \qquad \text{or} \qquad p = \text{C}.$$

This value of p in (9) gives

$$y = \text{C}x + n\sqrt{1 + \text{C}^2}.$$

From the other factor we have

$$p = \pm \frac{x}{\sqrt{n^2 - x^2}},$$

which, in (9), gives

$$y^2 + x^2 = n^2,$$

a result containing no arbitrary constant, which will be further explained in the following article.

Singular Solutions.

215. It has been seen, that many differential equations of the first order result from the elimination of a constant from the primitive equation and its immediate differential. Thus, let

$$f(x, y, \text{C}) = 0 \ldots\ldots\ldots (1),$$

be the primitive equation containing the variables x and y, and the constant C;

$$Pdx + Qdy = 0 \ldots\ldots\ldots(2),$$

is immediate differential equation; and

$$P'dx + Q'dy = 0 \ldots\ldots\ldots(3),$$

the result obtained by the elimination of C from (1) and (2). It may now be asked: May not such a function of x and y be substituted for C, that the result of the combination of equation (1), under this supposition, with its immediate differential, shall be the same as before? To answer this, let equation (1) be differentiated, x, y, and C being regarded as variables; we thus obtain

$$Pdx + Qdy + C'dC = 0 \ldots.(4).$$

Now, if $C'dC = 0$, it is plain that equation (4) will be the same as equation (2), and the result of the elimination of C between it and (1), will then be the same as equation (3).

If, then, for C in equation (1), we substitute the variable value deduced from the equation

$$C'dC = 0,$$

that equation will contain no arbitrary constant, and yet will be as much a primitive equation as any one containing the arbitrary constant.

Such results are termed *singular solutions*, inasmuch as they cannot possibly be obtained from the complete integral, Art. (160), by assigning to the arbitrary constant a particular value; the latter results being called *particular integrals*.

The equation $C'dC = 0$ can be satisfied, by making

$$dC = 0, \qquad \text{or} \qquad C' = 0.$$

The first gives $C =$ a constant, the particular values of which, when substituted in equation (1), give particular integrals.

The values of C deduced from $C' = 0$, if variable, will then give the only singular solutions.

To illustrate, let us resume the complete integral of equation (9), in the preceding article,

$$y = Cx + n\sqrt{1 + C^2} \ldots\ldots (5).$$

Differentiating with reference to C, we have

$$0 = xdC + \frac{nCdC}{\sqrt{1 + C^2}};$$

whence

$$x + \frac{nC}{\sqrt{1 + C^2}} = 0 \ldots\ldots (6),$$

and

$$x^2 + x^2C^2 = n^2C^2, \quad \text{or} \quad C = -\frac{x}{\sqrt{n^2 - x^2}};$$

the negative value of C being plainly the only one which will satisfy equation (6). Its substitution in (5) gives

$$y = -\frac{x^2}{\sqrt{n^2 - x^2}} + n\sqrt{\frac{n^2}{n^2 - x^2}}$$

$$y = \sqrt{n^2 - x^2}, \quad \text{or} \quad y^2 + x^2 = n^2,$$

the singular solution found in the preceding article.

Integration of Differential Equations of the Second Order.

216. Of these equations, which, in their most general form, contain $\frac{d^2y}{dx^2}$, $\frac{dy}{dx}$, y, x, and constants, we shall only discuss those particular cases which admit of integration.

I. The proposed equation may contain only $\frac{d^2y}{dx^2}$, x, and constants; in which case, solving it with reference to $\frac{d^2y}{dx^2}$, we have

$$\frac{d^2y}{dx^2} = f(x),$$

which may be integrated as in article (193).

217. II. It may contain only $\frac{d^2y}{dx^2}$, y, and constants. Solving the equation as before, we obtain

$$\frac{d^2y}{dx^2} = \text{Y}.$$

Multiplying by $2dy$,

$$\frac{2dy}{dx}\frac{d^2y}{dx} = 2\text{Y}dy,$$

and integrating,

$$\frac{dy^2}{dx^2} = \int 2\text{Y}dy + \text{C}, \qquad \text{or} \qquad \frac{dy}{dx} = \sqrt{2\int \text{Y}dy + \text{C}};$$

whence

$$dx = \frac{dy}{\sqrt{2\int \text{Y}dy + \text{C}}}, \qquad x = \int \frac{dy}{\sqrt{2\int \text{Y}dy + \text{C}}} + \text{C}'.$$

Examples.

1. If $$a^2 d^2 y + y dx^2 = 0,$$

$$\frac{d^2y}{dx^2} = -\frac{y}{a^2}, \qquad \frac{2dy}{dx}\frac{d^2y}{dx} = -\frac{2ydy}{a^2},$$

$$\frac{dy^2}{dx^2} = -\frac{y^2}{a^2} + C, \qquad \frac{dy}{dx} = \sqrt{C - \frac{y^2}{a^2}},$$

$$x = \int \frac{dy}{\sqrt{C - \frac{y^2}{a^2}}} + C',$$

which may be integrated as in example (5), article (162).

2. Let $$d^2 y \sqrt{ay} = dx^2.$$

218. III. The equation may contain only $\frac{d^2y}{dx^2}$, $\frac{dy}{dx}$, and constants, being expressed generally thus,

$$F\left(\frac{d^2y}{dx^2}, \frac{dy}{dx}\right) = 0 \ldots\ldots\ldots (1).$$

Make $\frac{dy}{dx} = p$; then $\frac{d^2y}{dx^2} = \frac{dp}{dx}$, and (1) becomes

$$F\left(\frac{dp}{dx}, p\right) = 0,$$

which is of the first order with reference to dp, and may be solved with reference to dx; whence

$$dx = F'(p)\,dp \ldots (2), \qquad x = \int F'(p)\,dp + C \ldots (3).$$

Multiplying (2) by p, we have

$$pdx = dy = pF'(p)\,dp, \qquad y = \int pF'(p)\,dp + C' \ldots (4).$$

Eliminating p from (3) and (4), we have the primitive equation between x, y, and the two arbitrary constants C and C'.

For an example, let

$$\frac{(dx^2 + dy^2)^{\frac{3}{2}}}{dxd^2y} = a, \qquad \text{or} \qquad \frac{dx(1 + p^2)^{\frac{3}{2}}}{dp} = a;$$

whence

$$dx = \frac{adp}{(1 + p^2)^{\frac{3}{2}}}, \qquad pdx = dy = \frac{apdp}{(1 + p^2)^{\frac{3}{2}}}.$$

Integrating the last two expressions, we have

$$x = C + \frac{ap}{\sqrt{1 + p^2}}, \qquad y = C' - \frac{a}{\sqrt{1 + p^2}},$$

and eliminating p,

$$(x - C)^2 + (y - C')^2 = a^2,$$

as was to be expected, since the proposed equation expresses constant radius of curvature.

219. IV. If the given equation does not contain y, it may be expressed

$$F\left(\frac{d^2y}{dx^2},\ \frac{dy}{dx},\ x\right) = 0, \qquad \text{or} \qquad F\left(\frac{dp}{dx},\ p,\ x\right) = 0,$$

which is of the first order with reference to dp. Its integral will give an equation of the form

$$f(p,\ x) = 0,$$

in which, p being replaced by $\frac{dy}{dx}$, and the result integrated, we shall have

$$f'(y,\ x) = 0,$$

with two arbitrary constants.

For an example, let

$$\frac{d^2y}{dx^2} = \frac{dy}{dx}\frac{1}{x},$$

or

$$\frac{dp}{dx} = \frac{p}{x}, \qquad \frac{dp}{p} = \frac{dx}{x},$$

$$lp = lx + C, \qquad p = C'x,$$

$$\frac{dy}{dx} = C'x, \qquad \text{and} \qquad y = \frac{C'x^2}{2} + C''.$$

220. V. If the given equation does not contain x, it may be expressed

$$F\left(\frac{d^2y}{dx^2},\ \frac{dy}{dx},\ y\right) = 0 \ldots\ldots\ldots (1).$$

Since $dy = pdx$,

$$dx = \frac{dy}{p}, \qquad \frac{d^2y}{dx^2} = \frac{dp}{dx} = \frac{pdp}{dy},$$

and equation (1) may be written

$$F\left(\frac{pdp}{dy}, p, y\right) = 0,$$

which is of the first order with reference to dp and dy. Its integral will then be expressed

$$F'(p, y) = 0, \qquad \text{or} \qquad F'\left(\frac{dy}{dx}, y\right) = 0,$$

and this may be treated as in case II., Art. (214).

221. VI. If the equation be of the form

$$\frac{d^2y}{dx^2} + X\frac{dy}{dx} + X'y = 0\ldots\ldots(1).$$

Make

$$y = e^{\int udx}\ldots\ldots\ldots\ldots(2);$$

then

$$\frac{dy}{dx} = ue^{\int udx}, \qquad \frac{d^2y}{dx^2} = e^{\int udx}\left(u^2 + \frac{du}{dx}\right).$$

These values in (1) give (since the common factor $e^{\int udx}$ disappears)

$$\frac{du}{dx} + u^2 + Xu + X' = 0,$$

which is of the first order with reference to du. After integration, the value of u being determined and substituted in (2), will give the required primitive equation,

$$y = e^{\int \int z dx}.$$

Integration of Differential Equations of Higher Orders than the Second.

222. Of these, it will also be sufficient for our purpose to discuss a few of the most simple cases.

I. Suppose the equation to contain only $\frac{d^n y}{dx^n}$, $\frac{d^{n-1} y}{dx^{n-1}}$, and constants; it may then be expressed,

$$F\left(\frac{d^n y}{dx^n}, \frac{d^{n-1} y}{dx^{n-1}}\right) = 0 \ldots\ldots (1).$$

Make

$$\frac{d^{n-1} y}{dx^{n-1}} = u; \qquad \text{then} \qquad \frac{d^n y}{dx^n} = \frac{du}{dx}.$$

These values in (1) give

$$F\left(\frac{du}{dx}, u\right) = 0,$$

which is of the first order; and its integral will give u in terms of x, or

$$u = X + C, \qquad \frac{d^{n-1} y}{dx^{n-1}} = X + C,$$

and finally,

$$y = \int^{n-1} (X + C)\, dx^{n-1}.$$

223. II. Suppose the equation expressed thus,

$$F\left(\frac{d^ny}{dx^n}, \frac{d^{n-2}y}{dx^{n-2}}\right) = 0 \ldots\ldots (1).$$

Make

$$\frac{d^{n-2}y}{dx^{n-2}} = u, \qquad \text{then} \qquad \frac{d^ny}{dx^n} = \frac{d^2u}{dx^2},$$

and equation (1) will become

$$F\left(\frac{d^2u}{dx^2}, u\right) = 0,$$

which may be integrated as in article (217), and the value of $u = f(x)$ determined; we shall then have

$$\frac{d^{n-2}y}{dx^{n-2}} = f(x), \qquad \text{and} \qquad y = \int^{n-2} f(x)\, dx^{n-2}.$$

224. III. Suppose the equation to be of the form

$$d^3y + Ad^2ydx + Bdydx^2 + Dydx^3 = 0 \ldots (1).$$

Make

$$y = e^u \ldots\ldots\ldots\ldots\ldots\ldots (2),$$

u being an arbitrary function of x; then

$$dy = e^u du, \qquad d^2y = e^u(d^2u + du^2),$$

$$d^3y = e^u(d^3u + 3dud^2u + du^3).$$

These values in (1) give

$$d^3u + 3dud^2u + du^3 + A(d^2u + du^2)dx$$
$$+ Bdudx^2 + Ddx^3 = 0 \ldots\ldots\ldots\ldots (3).$$

Since u, in equation (2), is arbitrary, let such a value be assigned to it, that its differential shall be constant; in which case

$$du = mdx, \qquad d^2u = 0, \qquad d^3u = 0.$$

Equation (3), under this supposition, reduces to

$$m^3 + Am^2 + Bm + D = 0 \ldots\ldots\ldots (4).$$

From this equation we may determine the value of the constant m. Denoting the three roots by

$$m, \qquad m', \qquad m'',$$

we have for du the three values

$$du = mdx, \qquad du = m'dx, \qquad au = m''dx;$$

whence

$$u = mx + C, \quad u = m'x + C', \quad u = m''x + C'',$$

and

$$y = e^{mx+C}, \qquad y = e^{m'x+C'}, \qquad y = e^{m''x+C''};$$

or, calling

$$e^C = C, \qquad e^{C'} = C', \qquad e^{C''} = C'',$$

$$y = Ce^{mx}, \qquad y = C'e^{m'x}, \qquad y = C''e^{m''x}.$$

But since these values of y each contain but one arbitrary constant, they must be particular cases of the general value of y, which must be of such a form that either of the above particular values can be deduced from it; that is,

$$y = Ce^{mx} + C'e^{m'x} + C''e^{m''x},$$

from which the first particular value is deduced by making C′ and $C'' = 0$; and in a similar way, the others.

If two of the roots m, m', m'', are equal, that is, if $m = m'$, we should have the equation

$$y = (C + C')e^{mx} + C''e^{m''x} = Ce^{mx} + C''e^{m''x},$$

containing but *two* arbitrary constants, $C + C'$ being denoted by C. It is not then general. But in this case, $y = Ce^{mx}$, being a particular value,

$$y = C'xe^{mx} \dots\dots\dots\dots (5)$$

will be another; for, differentiating it, we have

$$dy = C'e^{mx}(1 + mx)dx,$$

$$d^2y = C'e^{mx}(2m + m^2x)dx^2,$$

$$d^3y = C'e^{mx}(3m^2 + m^3x)dx^3,$$

and these, substituted in equation (1), give

$$(m^3 + Am^2 + Bm + D)x + (3m^2 + 2Am + B) = 0 \dots (6).$$

But the coefficient of x is the same as the first member of equation (4), which has two roots equal to m; and $3m^2 + 2Am + B$ is its first derived polynomial, which, when placed equal to 0, must have one root equal to m (see Algebra); hence both terms of (6) are 0, and $y = C'xe^{mx}$ satisfies the given differential equation, and must therefore be a particular value of the general one,

$$y = Ce^{mx} + C'xe^{mx} + C''e^{m''x}.$$

If $m = m' = m''$, it may be shown also by trial, as above, that

$$y = C''x^2e^{mx}$$

is a particular value; whence the general value must be

$$y = e^{mx}(C + C'x + C''x^2).$$

Two of the roots may be imaginary; but, as the discussion in this case is quite complicated, and of little value to the student, we omit it.

To illustrate the above, let

$$d^3y + 2d^2ydx - dydx^2 - 2ydx^3 = 0.$$

Comparing this with equation (1), we have

$$A = 2, \qquad B = -1, \qquad D = -2;$$

and equation (4) becomes

$$m^3 + 2m^2 - m - 2 = 0;$$

whence the three values of m are

$$-2, \qquad 1, \qquad \text{and} \qquad -1,$$

and the general value of y is

$$y = Ce^{-2x} + C'e^x + C''e^{-x}.$$

225. It is plain that the preceding principles can readily be extended to the general equation

$$d^ny + Ad^{n-1}ydx + Bd^{n-2}ydx^2 \ldots\ldots + Mydx^n = 0,$$

and that the general value of y will be

$$y = Ce^{mx} + C'e^{m'x} + C''e^{m''x} + \&c.$$

226. If the equation be

$$d^3y + Xd^2ydx + X'dydx^2 + X''ydx^3 = 0 \ldots (1),$$

in which X, &c., are functions of x, the difficulty of integration is much increased. If, however, we know three particular values of y; Cy', $C'y''$, $C''y'''$, each of which will satisfy the given equation, then the general value of y will equal their sum, that is,

$$y = Cy' + C'y'' + C''y''' \ldots\ldots\ldots\ldots (2).$$

To verify this, let equation (2) be differentiated three times, and the proper values substituted in (1); we shall thus obtain

$$\left.\begin{array}{l} C(d^3y' + Xd^2y'dx + X'dy'dx^2 + X''y'dx^3) \\ + C'(d^3y'' + Xd^2y''dx + X'dy''dx^2 + X''y''dx^3) \\ + C''(d^3y''' + Xd^2y'''dx + X'dy'''dx^2 + X''y'''dx^3) \end{array}\right\} = 0,$$

which is satisfied, since each of the three terms is, by hypothesis, equal to 0.

227. The above demonstration can be generalized, and a similar result obtained for the equation

$$d^ny + Xd^{n-1}ydx + \ldots\ldots\ldots yX^{(n-1)\prime}dx^n = 0.$$

This, and the equations discussed in the three preceding articles, belong to the class termed *linear*. See note to article (205).

Integration of Partial Differential Equations of the First Order.

228. A partial differential equation of the first order, derived from an equation between the three variables z, y, and x, z being regarded as a function of x and y, contains, in its most general form, the three variables, the two partial differential coefficients, $\frac{dz}{dx}$ and $\frac{dz}{dy}$, and constants. Without attempting to discuss the most general, we will confine ourselves to a few of the most simple cases.

I. If the equation contains but one partial differential coefficient and the two independent variables, that is, if

$$\frac{dz}{dx} = \mathrm{P},$$

P being a function of x and y; we integrate at once as in article (195). For example, if

$$\frac{dz}{dx} = \frac{x}{\sqrt{x^2 + y^2}}, \qquad z = \sqrt{x^2 + y^2} + \mathrm{Y}.$$

229. II. Let the equation be

$$\frac{dz}{dx} = \mathrm{R},$$

R being a function of the three variables. Since the partial differential coefficient has been obtained under the supposition that y is constant, the proposed equation may be regarded as a differential equation between z and x, and may be integrated as in article (201), taking care to add an arbitrary function of y.

Examples.

1. Let $$\frac{dz}{dx} = \frac{\sqrt{y^2 - z^2}}{z}x.$$

By the separation of the variables, we have

$$xdx = \frac{zdz}{\sqrt{y^2 - z^2}},$$

and by integration,

$$\frac{x^2}{2} = - \sqrt{y^2 - z^2} + \varphi(y).$$

2. Let $$\frac{dz}{dx} = \frac{y^2 + z^2}{y^2 + x^2}.$$

230. III. Let the equation be

$$\mathrm{M}\frac{dz}{dy} + \mathrm{N}\frac{dz}{dx} = 0,$$

M and N being functions of x and y.

Solving the equation with reference to $\frac{dz}{dy}$, we have

$$\frac{dz}{dy} = -\frac{\mathrm{N}}{\mathrm{M}}\frac{dz}{dx}.$$

But, since z is a function of x and y,

$$dz = \frac{dz}{dx}dx + \frac{dz}{dy}dy;$$

or, by the substitution of the value of $\frac{dz}{dy}$,

$$dz = \frac{dz}{dx}(dx - \frac{N}{M}dy) = \frac{dz}{dx}\left(\frac{Mdx - Ndy}{M}\right)\ldots(1).$$

If S be the factor which will make $Mdx - Ndy$ integrable, we may write

$$S(Mdx - Ndy) = du,$$

which, in (1), gives

$$dz = \frac{1}{SM}\frac{dz}{dx}du,$$

to satisfy which, it is only necessary that $\frac{1}{SM}\frac{dz}{dx} = F(u)$; whence

$$dz = F(u)\,du, \qquad z = \varphi(u),$$

the form of this function being arbitrary.

Examples.

1. If $$x\frac{dz}{dy} - y\frac{dz}{dx} = 0,$$

$$Mdx - Ndy = xdx + ydy,$$

which is made integrable by the factor 2, and we have

$$x^2 + y^2 = u, \quad \text{and} \quad z = \varphi(x^2 + y^2),$$

which is the general equation of a surface of revolution.

2. If
$$y\frac{dz}{dy} + x\frac{dz}{dx} = 0,$$

$$\text{M}dx - \text{N}dy = ydx - xdy,$$

which may be integrated by the aid of the factor $\frac{1}{y^2}$; whence

$$\frac{x}{y} = u, \quad \text{and} \quad z = \varphi\left(\frac{x}{y}\right).$$

Application of the Calculus to the Determination of Curves with Particular Properties.

231. By means of the preceding principles, we are often enabled to deduce the equation of a curve which shall possess a particular property.

I. Let it be required to find a curve whose subnormal shall be constant. The constant being denoted by a, we place the general expression for the subnormal, Art. (85), equal to a, and have

$$y\frac{dy}{dx} = a, \quad \text{whence} \quad ydy = adx;$$

and integrating,

$$\frac{y^2}{2} = ax + \text{C}, \quad y^2 = 2ax + 2\text{C},$$

the equation of a parabola.

II. Find the equation of a curve whose subtangent is constant. Place

$$y\frac{dx}{dy} = a, \quad \text{whence} \quad a\frac{dy}{y} = dx.$$

Integrating,

$$aly = x, \qquad \text{or} \qquad x = \log y + C,$$

the equation of a logarithmic curve, the modulus of the system being a.

III. Let problem I. be generalized, and let it be required to find a curve whose subnormal shall be a given function of the abscissa, denoted by X. Then

$$y\frac{dy}{dx} = X, \qquad ydy = Xdx,$$

$$y^2 = 2\int Xdx.$$

As a particular case, let $X = \frac{x^2}{a}$. Then

$$y^2 = 2\int \frac{x^2}{a} dx = \frac{2}{3}\frac{x^3}{a} + C.$$

IV. Let problem II. be generalized in like manner. Then

$$y\frac{dx}{dy} = X, \qquad \frac{dx}{X} = \frac{dy}{y}, \qquad ly = \int \frac{dx}{X}.$$

As a particular case, let $X = 2x$. Then

$$ly = \int \frac{dx}{2x} = \frac{1}{2} lx + C,$$

$$2ly = lx + 2C, \qquad ly^2 = lx + 2C;$$

or, denoting $2C$ by lC',

$$ly^2 = lC'x, \qquad y^2 = C'x,$$

the equation of a common parabola.

V. To find a curve whose normal is constant. Place the general expression for the normal, Art. (85),

$$y\sqrt{1 + \frac{dy^2}{dx^2}} = a,$$

whence

$$y^2 + y^2\frac{dy^2}{dx^2} = a^2, \qquad \frac{ydy}{(a^2 - y^2)^{\frac{1}{2}}} = dx;$$

and by integration,

$$-(a^2 - y^2)^{\frac{1}{2}} = x + C, \quad \text{or} \quad a^2 - y^2 = (x + C)^2,$$

the equation of a circle.

VI. The curve whose tangent is constant may also be found by placing

$$y\sqrt{1 + \frac{dx^2}{dy^2}} = a,$$

and this problem and the preceding may be generalized as in problems III. and IV.

VII. Required to find a curve, such that its normal shall be a mean proportional between a given line and the sum of its abscissa and subnormal.

We have at once from the conditions,

$$2a\left(x + y\frac{dy}{dx}\right) = y^2\left(1 + \frac{dy^2}{dx^2}\right),$$

$2a$ denoting the given line. Solving this with reference to $\frac{dy}{dx}$, Art. (213), we have, for the first value,

$$\frac{dy}{dx} = \frac{a}{y} + \left(\frac{a^2}{y^2} + \frac{2ax}{y^2} - 1\right)^{\frac{1}{2}};$$

whence

$$\frac{2adx - 2ydy}{2(a^2 + 2ax - y^2)^{\frac{1}{2}}} = - dx.$$

The integral of the first member is evidently the radical in the denominator, Art. (25), and we have

$$(a^2 + 2ax - y^2)^{\frac{1}{2}} = - x + C,$$

or

$$a^2 + 2ax - y^2 = (C - x)^2,$$

the equation of a circle.

232. Let it be required to find a curve which shall intersect, at a given angle, a class of curves whose equation contains but one arbitrary constant.

Let the general equation of the class of curves be

$$y' = f(a, x') \ldots\ldots\ldots\ldots (1),$$

a being the only arbitrary constant; by assigning values to which, in succession, the particular curves are determined; and let x and y denote the co-ordinates of the required curve. Then if T denote the tangent of the angle at which this curve intersects each particular curve, we have

$$T = \frac{p - p'}{1 + pp'} \ldots\ldots\ldots (2),$$

in which p and p' are the tangents of the angles which the tangents to the curves, at their point of intersection, make with the axis of X.

If the given equation be differentiated, and the expression for p' be found and substituted in (2), and then the result combined with (1), and a be eliminated, the final equation will belong to no one of the particular curves more than to another. If, in this equation, x' be made equal to x, and $y' = y$, since for the point of intersection of the curves these variables are equal, we shall have the differential equation of the required curve.

1. Let the equation of the class be

$$y' = ax' \ldots\ldots\ldots (3), \qquad \text{whence} \qquad \frac{dy'}{dx'} = a = p';$$

and let the angle be 45°, in which case $\text{T} = 1$.

Substituting these in (2), and reducing, we have

$$\frac{dy}{dx} - a = 1 + a\frac{dy}{dx}.$$

Eliminating a by substituting its value taken from equation (3), making at the same time $x' = x$, and $y' = y$, we have

$$\frac{dy}{dx} - \frac{y}{x} = 1 + \frac{y}{x}\frac{dy}{dx}.$$

This, being homogeneous, may be integrated as in Art. (203) or Art. (211), and we shall obtain

$$l\frac{\sqrt{x^2 + y^2}}{\text{C}} = \tan^{-1}\frac{y}{x}.$$

If in this we put for x, $r \cos v$, and for y, $r \sin v$, by which the reference is changed to the system of polar co-ordinates, w have

$$\frac{y}{x} = \tan v, \qquad \sqrt{x^2 + y^2} = r,$$

$$l\frac{r}{C} = \tan^{-1}(\tan v), \quad \text{or} \quad l\frac{r}{C} = v,$$

the equation of a logarithmic spiral, Art. (143).

2. Take the parabolas given by the equation

$$y'^2 = mx' \ldots\ldots (4), \qquad \text{whence} \qquad \frac{dy'}{dx'} = \frac{m}{2y'} = p',$$

and let it be required to find the curve which cuts these parabolas at right angles. In this case $T = \infty$, and we must have

$$1 + pp' = 0, \qquad \text{or} \qquad 1 + \frac{dy}{dx}\frac{dy'}{dx'} = 0.$$

Substituting the above value of p', we have

$$1 + \frac{mp}{2y'} = 0.$$

Eliminating m by equation (4), and making $y' = y$,

$$1 + \frac{y}{2x}\frac{dy}{dx} = 0, \qquad \text{or} \qquad 2xdx + ydy = 0.$$

Integrating,

$$x^2 + \frac{y^2}{2} = C, \qquad \text{or} \qquad 2x^2 + y^2 = 2C,$$

the equation of an ellipse.

Rectification of Curves.

233. *The rectification of a curve* is any operation by which the measure of its length is obtained.

In article (90), we have shown how to find an expression for the differential of an arc of a plane curve, in terms of either variable and its differential. If this expression can be integrated, we can, by its integration, obtain an expression for the curve itself. From this results the following simple rule for the rectification of any plane curve: *Deduce, as in Art.* (90), *an expression for the differential of the arc, in terms of either variable and its differential, and integrate the result.* We shall thus obtain an expression for *an indefinite portion of the arc.* For the length of a definite portion, take the integral between the limits designated by those values of the variable which correspond to its extremities, Art. (160), and the numerical value of this expression will be the required measure.

234. The curves represented by the general equation

$$y^n = px^m,$$

in which m and n are entire and positive, are called parabolas. This equation can be written

$$y = p^{\frac{1}{n}}x^{\frac{m}{n}} = p'x^r \ldots\ldots\ldots (1).$$

By differentiation, we have

$$dy = rp'x^{r-1}dx.$$

By substituting this in the expression

$$dz = \sqrt{dx^2 + dy^2},$$

and indicating the integration, we have

$$z = \int \sqrt{dx^2 + dy^2} = \int dx\,(1 + r^2p'^2x^{2r-2})^{\frac{1}{2}}.$$

This admits of an exact integral, when either

$$\frac{1}{2r-2}, \qquad \text{or} \qquad -\left(\frac{1}{2r-2} + \frac{1}{2}\right),$$

is equal to a whole number, Art. (176); and a general expression for the length of the curves may thus be found in terms of x.

If, in equation (1), we make $r = \frac{3}{2}$, we have

$$y = p'x^{\frac{3}{2}}, \qquad \text{or} \qquad y^2 = p'^2x^3,$$

which is the equation of a cubic parabola. In this case,

$$z = \int dx\,(1 + \frac{9}{4}p'^2x)^{\frac{1}{2}} = \frac{8}{27p'^2}(1 + \frac{9}{4}p'^2x)^{\frac{3}{2}} + C.$$

If we wish the length from that point whose abscissa is a, to that whose abscissa is b, we take the integral between the limits a and b.

Let us, however, estimate the arc from the vertex, or suppose the origin of the integral to be at the point where $x = 0$, Art. (160); we then have

$$0 = \frac{8}{27p'^2} + C, \qquad \text{or} \qquad C = -\frac{8}{27p'^2};$$

whence, denoting this particular integral by z',

$$z' = \frac{8}{27p'^2}[(1 + \frac{9}{4}p'^2x)^{\frac{3}{2}} - 1],$$

for the length of any arc from the vertex to the point whose abscissa is x.

If $r = \frac{1}{2}$, we have

$$y = p'x^{\frac{1}{2}}, \qquad \text{or} \qquad y^2 = p'^2x,$$

the equation of the common parabola. In this case,

$$z = \int dx\,(1 + \frac{1}{4}p'^2x^{-1})^{\frac{1}{2}} = \sqrt{\frac{4x + p'^2}{4x}}dx,$$

which may be made rational, and integrated as in Art. (172).

235. The length of the common parabola may also be determined in terms of y. By differentiating the equation

$$y^2 = 2px,$$

we obtain

$$2ydy = 2pdx, \qquad dx = \frac{ydy}{p}.$$

This value in the expression $z = \int \sqrt{dx^2 + dy^2}$, gives

$$z = \int dy \sqrt{1 + \frac{y^2}{p^2}} = \frac{1}{p}\int dy\,(p^2 + y^2)^{\frac{1}{2}},$$

which, by formula **B**, may be reduced to

$$z = \frac{y\sqrt{p^2 + y^2}}{2p} + \frac{p}{2}\int \frac{dy}{\sqrt{p^2 + y^2}}.$$

But

$$\int \frac{dy}{\sqrt{p^2 + y^2}} = l(\sqrt{p^2 + y^2} + y) \ldots\ldots \text{Art. (173)};$$

hence

$$z = \frac{y\sqrt{p^2 + y^2}}{2p} + \frac{p}{2} l(\sqrt{p^2 + y^2} + y) + \text{C}.$$

If we estimate the arc from the vertex, where $y = 0$, we have

$$0 = \frac{p}{2} lp + \text{C}, \qquad \text{or} \qquad \text{C} = -\frac{p}{2} lp;$$

and finally, denoting the particular integral by z',

$$z' = \frac{y\sqrt{p^2 + y^2}}{2p} + \frac{p}{2}[l(\sqrt{p^2 + y^2} + y) - lp].$$

236. For the arc of the circle, we have, Art. (90),

$$z = \text{R}\int \frac{dx}{\sqrt{\text{R}^2 - x^2}} = \text{R} \sin^{-1}\frac{x}{\text{R}},$$

which *can be expressed by a series*, and the length of z determined approximately.

Differentiating the equation of the ellipse, we deduce

$$dy = -\frac{b^2 x}{a^2 y} dx;$$

whence

$$z = \int dx \sqrt{1 + \frac{b^4 x^2}{a^4 y^2}} = \frac{1}{a}\int \frac{dx\sqrt{a^4 - (a^2 - b^2)x^2}}{\sqrt{a^2 - x^2}},$$

which can only be expressed by a series.

237. The differential equation of the cycloid, Art. (131), is

$$dx = \frac{ydy}{\sqrt{2ry - y^2}}.$$

By the substitution of this value of dx, we obtain

$$z = \int \sqrt{dx^2 + dy^2} = \int dy \sqrt{\frac{2ry}{2ry - y^2}} = \sqrt{2r} \int dy\, (2r - y)^{-\frac{1}{2}};$$

whence, article (158),

$$z = -\, 2\sqrt{2r}\,(2r - y)^{\frac{1}{2}} + \mathrm{C} = -\, 2\sqrt{2r\,(2r - y)} + \mathrm{C}.$$

If we estimate the arc from the point D, where $y = 2r$, we have

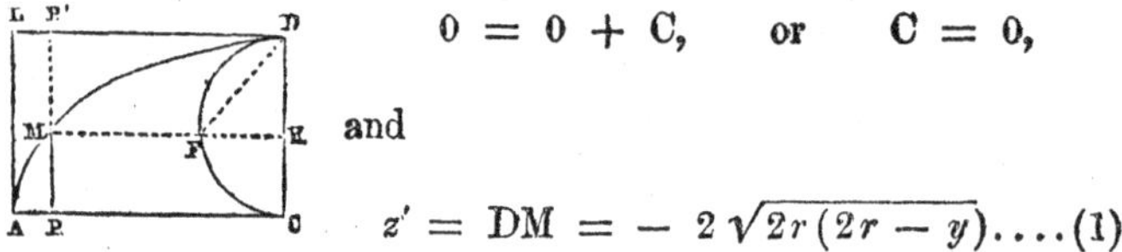

$$0 = 0 + \mathrm{C}, \quad \text{or} \quad \mathrm{C} = 0,$$

and

$$z' = \mathrm{DM} = -\, 2\sqrt{2r\,(2r - y)} \ldots . (1).$$

From the figure we see that

$$\mathrm{DF} = \sqrt{\mathrm{DC} \times \mathrm{DH}} = \sqrt{2r\,(2r - y)};$$

hence

$$\mathrm{DM} = -\, 2\mathrm{DF},$$

or the arc is equal to *twice the corresponding chord of the generating circle.*

If in equation (1) we make $y = 0$, and denote the definite integral by z'', we have

$$z'' = \mathrm{DMA} = -\, 4r = -\, 2\mathrm{DC},$$

as in article (135).

238. For the rectification of the spirals, we take the expression in article (138),

$$dz = \sqrt{dr^2 + r^2dv^2}.$$

By differentiating the general equation $r = av^n$, we deduce

$$dr^2 = n^2a^2v^{2n-2}dv^2;$$

whence, by substitution, &c.,

$$z = \int av^{n-1}dv\sqrt{n^2 + v^2}\ldots\ldots(1),$$

For the spiral of Archimedes, Art. (140), $n = 1$, $a = \frac{1}{2\pi}$, and the expression becomes

$$z = \frac{1}{2\pi}\int dv\sqrt{1 + v^2},$$

and the particular integral estimated from the pole may be obtained by placing 1 for p, and v for y, in the expression for z', in Art. (235); whence, after multiplying by $\frac{1}{2\pi}$,

$$z' = \frac{1}{4\pi}(v\sqrt{1 + v^2}) + l(\sqrt{1 + v^2} + v).$$

For the hyperbolic spiral $n = -1$, and expression (1) becomes

$$z = a\int v^{-2}dv\sqrt{1 + v^2}.$$

For the logarithmic spiral, when $M = 1$, we have

$$v = lr, \qquad dv = \frac{dr}{r},$$

$$z = \int dr\sqrt{2} = r\sqrt{2} + C;$$

or, estimating the arc from the pole, where $r = 0$, we have

$$z' = r\sqrt{2},$$

or *the diagonal of the square upon the radius vector.*

Quadrature of Curves.

239. *The quadrature of a curve* is the operation by which the measure of the area limited by it, is determined.

To determine the area limited by the curve and either of the co-ordinate axes, *we find, as in article* (92), *an expression for the differential of the area in terms of one variable and its differential, and integrate this. The result will be a general expression for an indefinite portion of the area.* For a definite portion, we take the integral between the limits designated by those values of the variable belonging to the extremities of the limiting curve. The numerical value of this will be the required measure.

240. The value of y, taken from the general equation of parabolas, Art. (234), is

$$y = p'x^r \ldots\ldots\ldots\ldots (1),$$

which, in the formula $ds = ydx$, gives

$$s = \int p'x^r dx = \frac{p'x^{r+1}}{r+1} + C.$$

If we estimate the area from the origin, where $x = 0$, we have

$$C = 0;$$

whence

$$s' = \frac{p'x^{r+1}}{r+1} = \frac{yx}{r+1},$$

that is, the area of a portion of a parabola, included between the curve, the axis of X, and any assumed ordinate, is equal to the rectangle of the ordinate and corresponding abscissa, divided by $r + 1$. Hence this portion of any parabola is always commensurable with this rectangle.

The same result may be obtained otherwise, thus: The value of from (1) is

$$x = \frac{y^{\frac{1}{r}}}{p'^{\frac{1}{r}}}, \qquad \text{whence} \qquad dx = \frac{y^{\frac{1}{r}-1}dy}{rp'^{\frac{1}{r}}},$$

and this, in the formula, gives

$$s = \int \frac{y^{\frac{1}{r}}dy}{rp'^{\frac{1}{r}}} = \frac{1}{r+1}\frac{1}{p'^{\frac{1}{r}}}y^{\frac{1}{r}+1} = \frac{yx}{r+1} + C,$$

as before.

For the common parabola, we have $r = \frac{1}{2}$; whence

$$s' = \frac{yx}{\frac{1}{2}+1} = \frac{2}{3}xy.$$

For the cubic parabola, $r = \frac{3}{2}$; whence

$$s' = \frac{2}{5}xy.$$

241. The value of y taken from the equation of the ellipse referred to its centre and axes, is

$$y = \frac{b}{a}\sqrt{a^2 - x^2};$$

hence

$$s = \frac{b}{a}\int(a^2 - x^2)^{\frac{1}{2}}dx.$$

By formula **B**, we have

$$\int(a^2 - x^2)^{\frac{1}{2}}dx = \frac{1}{2}x(a^2 - x^2)^{\frac{1}{2}} + \frac{1}{2}a^2\int dx\,(a^2 - x^2)^{-\frac{1}{2}}.$$

But

$$\int dx\,(a^2 - x^2)^{-\frac{1}{2}} = \int\frac{dx}{\sqrt{a^2 - x^2}} = \sin^{-1}\frac{x}{a} + \text{C};$$

whence, finally,

$$s = \frac{b}{2a}x\sqrt{a^2 - x^2} + \frac{ab}{2}\sin^{-1}\frac{x}{a} + \text{C}.$$

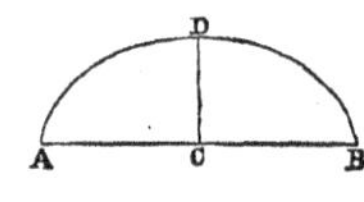

Taking the area between the limits

$$x = 0, \quad \text{and} \quad x = a,$$

we have

for $x = 0$, $$s = \frac{ab}{2}\sin^{-1}0 + \text{C} = \text{C};$$

for $x = a$, $$s = \frac{ab}{2}\sin^{-1}1 + \text{C} = \frac{ab}{2}\frac{\pi}{2} + \text{C};$$

and for the difference, or definite integral,

$$s'' = \frac{ab}{4}\pi = \text{CDB} = \frac{1}{4}\text{th of the ellipse};$$

hence the entire area is πab.

If $a = b$, the ellipse becomes a circle, of which a is the radius; whence the area of the circle is

$$\pi a^2 = \pi (\text{radius})^2.$$

The same result may be obtained by taking the value

$$y = \sqrt{2Rx - x^2};$$

whence

$$s = \int dx \sqrt{2Rx - x^2};$$

for the area of an indefinite portion of the circle.

242. In order to find an expression for the area of a portion of the hyperbola, it will be best to take its equation when referred to the centre and asymptotes,

$$xy = m;$$

and, since the asymptotes are oblique to each other, we must use the formula deduced in article (92),

$$ds = \sin\beta\, y dx,$$

β being the angle included by the asymptotes.

The value $$y = \frac{m}{x}$$

being substituted in the formula, gives

$$ds = \sin\beta \frac{mdx}{x}; \qquad \text{whence} \qquad s = \sin\beta\, mlx + C.$$

If we call the distance $CB = 1$, and estimate the area from the ordinate AB, for which $x = 1$, we have

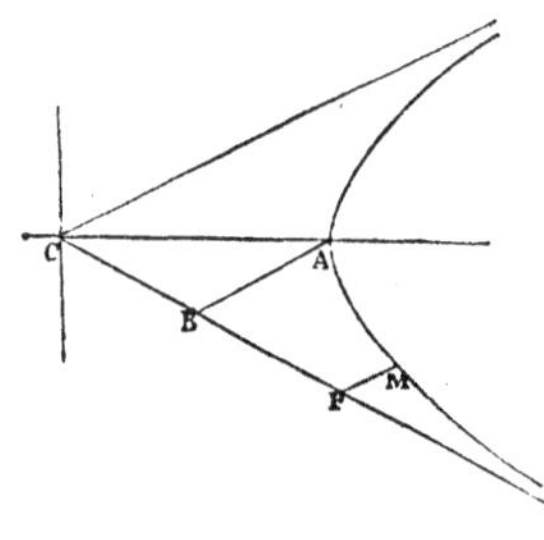

$$m = 1, \quad \text{and} \quad C = 0;$$

whence

$$s' = \sin\beta\, lx;$$

or, since $\sin\beta$ may be regarded as the modulus of a new system of logarithms, we have

$$s' = \log x;$$

or, *the area between the curve and asymptote, estimated from the ordinate of the vertex, is equal to the logarithm of the abscissa of its extreme point, taken in a system whose modulus is the sine of the angle made by the asymptotes.*

243. The value of dx taken from the differential equation of the cycloid, and substituted in the expression $s = \int ydx$, gives

$$s = \int \frac{y^2 dy}{\sqrt{2ry - y^2}},$$

which can be reduced by formula **E**, and finally integrated.

A more simple method, however, is to obtain directly the area ALD. If we denote $P'M = 2r - y$ by z, we shall have

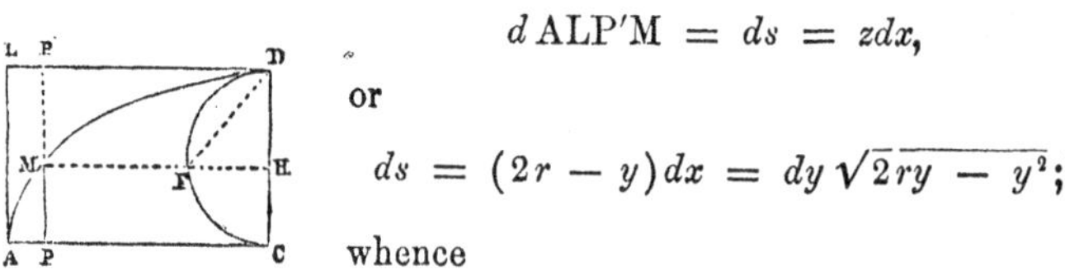

$$d\,ALP'M = ds = zdx,$$

or

$$ds = (2r - y)dx = dy\sqrt{2ry - y^2};$$

whence

$$s = \int dy \sqrt{2ry - y^2}.$$

But this is evidently the area of a portion of a circle whose radius is r, and abscissa y, Art. (241); that is, the area of the segment CFH. If we estimate these areas, the first from AL, and the second from the point C, they will both be 0, when $y = 0$; the arbitrary constant to be added in each case will then be 0, and we have

$$\text{ALP'M} = \text{CFH},$$

and when $y = 2r$,

$$\text{ALD} = \text{CFD} = \frac{\pi r^2}{2}.$$

But the area of the rectangle

$$\text{ALDC} = \text{AC} \times \text{CD} = \pi r . 2r = 2\pi r^2;$$

hence

$$\text{area AMDC} = \text{ALDC} - \text{ALD} = \frac{3}{2}\pi r^2,$$

double of which, or the area included between one branch of the cycloid and its base, *is equal to three times the area of the generating circle.*

From this we see, also, that the area included between one branch of the cycloid and its base, is equal to *three-fourths of the rectangle described upon the base and axis.*

244. For the logarithmic curve

$$y = \log x;$$

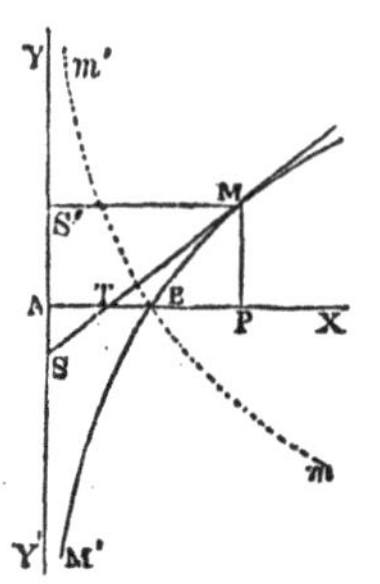

hence $$s = \int \log x \, dx,$$

or

$$s = x \log x - \mathrm{M}x + \mathrm{C} \ldots . \text{Art. (169)}.$$

M being the modulus.

If we estimate from the point B, where $x = 1$, we have

$$0 = -\mathrm{M} + \mathrm{C}, \qquad \mathrm{C} = \mathrm{M},$$

and

$$s' = x \log x - \mathrm{M}x + \mathrm{M}.$$

If we take the area included between the curve and axis of Y,

$$s = \int x dy = \int x \mathrm{M} \frac{dx}{x} = \mathrm{M}x + \mathrm{C},$$

or, estimating from the line AB, for which $x = 1$,

$$\mathrm{C} = -\mathrm{M}; \quad \text{whence} \quad s' = \mathrm{M}(x - 1).$$

If $x = 0$, we have $s'' = -\mathrm{M} =$ area Y'ABM'.

If $x = 2$, " $s'' = \mathrm{M} =$ area ABMS'.

245. The curve given by the equation

$$y^2 = \frac{1}{x},$$

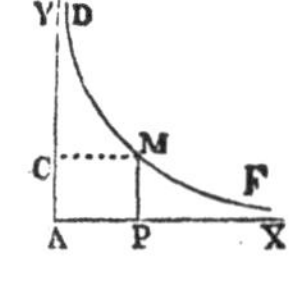

to which, as in the figure, the axes of co-ordinates are asymptotes, presents a similar case.

By differentiation, we obtain

$$dx = -\frac{2dy}{y^3};$$

whence

$$s = -\int\frac{2dy}{y^2} = \frac{2}{y} + C.$$

Estimating the area from the line AY, where $y = \infty$, we have

$$0 = \frac{2}{\infty} + C, \qquad C = 0,$$

and

$$s' = \frac{2}{y}.$$

By making $\quad y = 1 = MP, \quad$ we have

$$s'' = 2 = APMD;$$

that is, the area APMD is finite, and equal to twice the square APMC, although the curve does not touch the axis of Y at a finite distance.

If we take the area between the limits $y = 1$, and $y = 0$, we have

$$\text{area FMPX} = \frac{2}{0} - 2 = \infty.$$

246. For the quadrature of spirals, we take

$$ds = \frac{r^2dv}{2}\ldots\text{Art. (138)}, \qquad \text{or} \qquad s = \int\frac{r^2dv}{2}\ldots(1).$$

The value of r^2 taken from the general equation of spirals, Art. (139), is $r^2 = a^2v^{2n}$. This, substituted in formula (1), gives

$$s = \int\frac{a^2v^{2n}dv}{2} = \frac{a^2v^{2n+1}}{4n+2} + C.$$

Estimating the area from the pole, where $v = 0$ when n is positive, and ∞ when n is negative, we have, in all cases except when n is negative and numerically equal to or less than $\frac{1}{2}$, $C = 0$, and

$$s' = \frac{a^2 v^{2n+1}}{4n + 2}.$$

For the spiral of Archimedes, $n = 1$, and $a = \frac{1}{2\pi}$; whence

$$s' = \frac{v^3}{24\pi^2}.$$

If in this we make $v = 2\pi$, we have

$$s'' = \frac{\pi}{3},$$

which is the area PMA included within the first spire, or that described by one revolution of the radius vector. Since $PA = 1$, π represents the area of the circle PA; hence

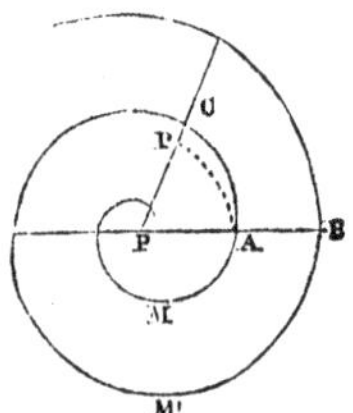

$$\text{area PMA} = \frac{1}{3} \text{ of the circle PA}$$

If $v = 2(2\pi)$,

we have
$$s'' = \frac{(4\pi)^3}{24\pi^2} = \frac{8}{3}\pi,$$

which is the whole area described by the radius vector during two revolutions. But it is plain that, during the second revolution, the part PMA will be described a second time; hence, to obtain the area PAM′B, we must subtract that described during the first revolution; we then have

$$\text{PAM'B} = \frac{8}{3}\pi - \frac{1}{3}\pi = \frac{7}{3}\pi;$$

and in general it will be seen, that by each revolution of the radius vector, the area before described will be increased by the area from the pole out to the last spire; hence, to obtain the area from the pole out to the mth spire; from the whole area described during m revolutions, take the area described during $m - 1$ revolutions; or take the integral between the limits

$$v = (m - 1)2\pi, \qquad \text{and} \qquad v = m2\pi,$$

which gives

$$\frac{(m2\pi)^3}{24\pi^2} - \frac{[(m-1)2\pi]^3}{24\pi^2} = \frac{m^3 - (m-1)^3}{3}\pi.$$

The area terminated by the $(m + 1)$th spire is then

$$\frac{(m+1)^3 - m^3}{3}\pi,$$

and the difference between the two expressions gives the area included between the mth and $(m + 1)$th spires, thus

$$\frac{(m+1)^3 - 2m^3 + (m-1)^3}{3}\pi = 2m\pi = m.2\pi.$$

If $m = 1$ in this expression, we have the area included between the first and second spire equal to 2π; hence, in general, the area between the mth and $(m + 1)$th spires *is equal to m times that included between the first and second.*

If the area PAC be required, AC being a portion of the second spire corresponding to the arc $\text{AD} = \frac{2\pi}{n'}$, we should have, for the whole area generated when the generating point has arrived at C, since $v = 2\pi + \frac{2\pi}{n'}$,

$$s'' = \frac{\left(2\pi + \frac{2\pi}{n'}\right)^3}{24\pi^2},$$

from which subtracting the area PMA, we have

$$\text{APC} = \frac{\left(2\pi + \frac{2\pi}{n'}\right)^3}{24\pi^2} - \frac{(2\pi)^3}{24\pi^2} = \frac{\pi}{n'}\left(1 + \frac{1}{n'} + \frac{1}{3n'^2}\right);$$

or, if we call AP (which has been regarded as unity), R,

$$\text{APC} = \frac{\pi}{n'}\left(1 + \frac{1}{n'} + \frac{1}{3n'^2}\right)\text{R}^2.$$

If $\text{AC} = \frac{1}{4}$ circumference $= \frac{2\pi}{4}$, then $n' = 4$, and

$$\text{APC} = \frac{\pi}{4}\left(1 + \frac{1}{4} + \frac{1}{48}\right)\text{R}^2.$$

For the hyperbolic spiral $n = -1$, and the general value of s' becomes

$$s' = -\frac{a^2}{2v},$$

which is infinite when $v = 0$. For the integral between the limits $v = b$ and $v = c$, we have

$$s'' = \frac{a^2}{2}\left(\frac{1}{b} - \frac{1}{c}\right).$$

In the logarithmic spiral, when $M = 1$,

$$v = lr, \qquad dv = \frac{dr}{r},$$

$$s = \int \frac{r^2 dv}{2} = \int \frac{r dr}{2} = \frac{r^2}{4} + C;$$

or, estimating from the pole, where $r = 0$ and $C = 0$, we have

$$s' = \frac{r^2}{4};$$

that is, *equal to one-fourth the square described upon the radius vector of the extreme point of the curve.*

Area of Curved Surfaces.

247. I. *Of surfaces of revolution.* In article (93), we have found, for the differential of the area of a surface of revolution, $du = 2\pi y \sqrt{dx^2 + dy^2}$; whence, for the indefinite area, we have

$$u = \int 2\pi y \sqrt{dx^2 + dy^2} \ldots\ldots\ldots\ldots (1),$$

the axis of X being the axis of revolution, and $\sqrt{dx^2 + dy^2}$ the differential of the arc of the generating curve.

The indefinite area of any particular surface will then be obtained *by deducing, as in Art.* (93), *the expression for the differential of the surface, in terms of one variable and its differential, and integrating the result.*

248. Let the line AC, by its revolution about AB, generate the surface of a right cone. The origin of co-ordinates being at A, the equation of AC is

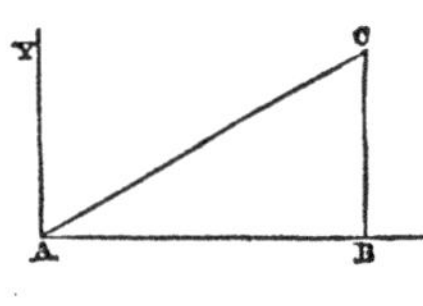

$$y = ax; \quad \text{whence} \quad dy = adx,$$

and

$$u = \int 2\pi a x dx \sqrt{a^2 + 1} = \pi a x^2 \sqrt{a^2 + 1} + C.$$

Estimating the area from the vertex, where $x = 0$, we have $C = 0$, and

$$u' = \pi a x^2 \sqrt{a^2 + 1}.$$

Making $x = AB = h$, we have the area of the cone whose altitude is h, and the radius of the base $BC = b$,

$$u'' = \pi a h^2 \sqrt{a^2 + 1};$$

or, since $a = \frac{b}{h}$,

$$u'' = \frac{2\pi b \sqrt{b^2 + h^2}}{2} = 2\pi b \frac{AC}{2};$$

that is, *the circumference of the base into half the side.*

249. From the equation of the circle, we have

$$y = \sqrt{2Rx - x^2}, \qquad dy = \frac{(R - x)\,dx}{y}.$$

The surface of the sphere is then

$$u = \int 2\pi y \sqrt{dx^2 + \frac{(\mathrm{R} - x)^2 dx^2}{y^2}} = \int 2\pi \mathrm{R} dx,$$

or

$$u = 2\pi \mathrm{R}x + \mathrm{C}.$$

Taking the area between the limits $x = 0$, and $x = 2\mathrm{R}$, we have

$$u'' = 4\pi\mathrm{R}^2 = \textit{four great circles.}$$

250. From the equation of the ellipse, we have

$$y = \frac{b}{a}\sqrt{a^2 - x^2}, \qquad dy = -\frac{b^2 x}{a^2 y}dx;$$

whence, for the area of the ellipsoid of revolution,

$$u = \int \frac{2\pi b}{a^2} dx \sqrt{a^4 - (a^2 - b^2)x^2}$$

$$= \frac{2\pi b}{a^2}\sqrt{a^2 - b^2}\int dx \sqrt{\frac{a^4}{a^2 - b^2} - x^2}$$

or, placing $\frac{2\pi b}{a^2}\sqrt{a^2 - b^2} = \mathrm{C}'$, and $\frac{a^4}{a^2 - b^2} = \mathrm{R}'^2$,

$$u = \mathrm{C}'\int dx \sqrt{\mathrm{R}'^2 - x^2}.$$

But $\int dx \sqrt{\mathrm{R}'^2 - x^2}$ = area of a circular segment whose radius is R', and abscissa x, Art. (92). Integrating this between the limits $x = 0$, and $x = \mathrm{CB} = a$, and calling the segment $\mathrm{CBFG} = \mathrm{D}$, we have

$$u'' = \mathrm{C}'\mathrm{D} = \frac{1}{2} \text{ area of ellipsoid.}$$

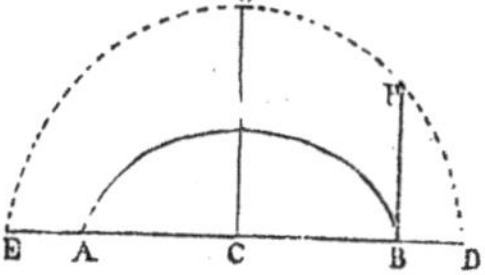

If $a = b$ in the primitive value of u, we shall have

$$u = \int 2\pi a dx = 2\pi a x + C,$$

for the surface of the circumscribing sphere.

Let the area of a paraboloid of revolution be determined.

251. By the substitution of the value of dx, Art. (237), in the general expression for u, we have for the surface generated by the revolution of a cycloid about its base,

$$u = 2\pi\sqrt{2r}\int y dy (2r - y)^{-\frac{1}{2}}.$$

Placing $2r - y = z$, and integrating as in Art. (159), we have

$$u = 2\pi\sqrt{2r}\left(-4r(2r - y)^{\frac{1}{2}} + \frac{2}{3}(2r - y)^{\frac{3}{2}}\right) + C.$$

Taking the area between the limits $y = 0$, and $y = 2r$, we have

$$u'' = \frac{32}{3}\pi r^2,$$

for one-half the surface. The whole is $\frac{64}{3}$ *the area of the generating circle.*

252. II. *Of curved surfaces generally.* In article (150), we have found, for a partial differential of the second order of a surface, the expression

$$d^2 u = dx dy \sqrt{1 + \left(\frac{dz}{dx}\right)^2 + \left(\frac{dz}{dy}\right)^2} \dots\dots (1).$$

If we differentiate the equation of any surface, first with reference to one independent variable, and then with reference to the other, and find expressions for the partial differential coefficients $\frac{dz}{dx}$ and $\frac{dz}{dy}$, in terms of x and y, and substitute in (1), and then integrate between proper limits, we shall obtain an expression for a definite portion of the surface.

For the sphere, we have

$$x^2 + y^2 + z^2 = R^2;$$

whence

$$\frac{dz}{dx} = -\frac{x}{z} = \frac{-x}{\sqrt{R^2 - x^2 - y^2}},$$

$$\frac{dz}{dy} = -\frac{y}{z} = \frac{-y}{\sqrt{R^2 - x^2 - y^2}},$$

$$\sqrt{1 + \left(\frac{dz}{dx}\right)^2 + \left(\frac{dz}{dy}\right)^2} = \frac{R}{\sqrt{R^2 - x^2 - y^2}},$$

and

$$u = \int^2 \frac{Rdxdy}{\sqrt{R^2 - x^2 - y^2}}.$$

Making $\sqrt{R^2 - y^2} = R'$, and integrating with reference to x, we have

$$u = \int Rdy \int \frac{dx}{\sqrt{R'^2 - x^2}} = \int Rdy \sin^{-1}\frac{x}{R'}$$

$$= \int Rdy\left(\sin^{-1}\frac{x}{\sqrt{R^2 - y^2}} + Y\right).$$

Taking the integral between the limits

$$x = 0, \qquad \text{and} \qquad x = ce' = \sqrt{R^2 - y^2},$$

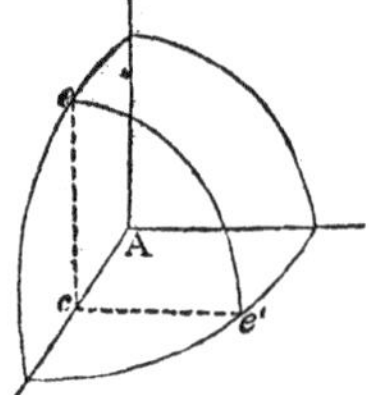

we have

$$u = \int R dy \frac{\pi}{2}.$$

Integrating again with reference to y, we have

$$u = \frac{R\pi}{2} y + C,$$

and between the limits $y = 0$, $y = R$,

$$u'' = \frac{\pi R^2}{2},$$

for one-eighth of the surface. The entire surface is then

$$4\pi R^2.$$

CUBATURE OF VOLUMES.

253. *The cubature* of a volume is any operation by which the measure of its contents is determined.

I. *Of volumes of revolution.* For the differential of a volume of revolution, we have found, Art. (94),

$$dv = \pi y^2 dx; \qquad \text{whence} \qquad v = \int \pi y^2 dx.$$

For the cubature of any particular volume, *we find, as in Art.* (94), *an expression for its differential, in terms of one variable and its differential, and then integrate; the result of the integration will be an expression for an indefinite portion of the volume.*

254. Let the rectangle ABCD revolve about AB and generate a right cylinder. The origin of co-ordinates being at A, the equation of DC will be

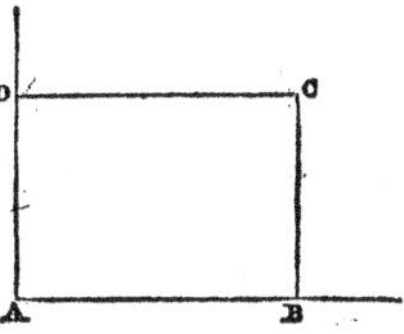

$$y = AD = b,$$

then

$$v = \int \pi y^2 dx = \int \pi b^2 dx = \pi b^2 x + C.$$

Taking this between the limits $x = 0$, and $x = AB = h$, we have

$$v'' = \pi b^2 h = \textit{the base into the altitude.}$$

255. The equation of the ellipse gives

$$y^2 = \frac{b^2}{a^2}(a^2 - x^2);$$

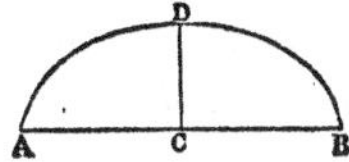

whence, for the ellipsoid of revolution,

$$v = \int \pi \frac{b^2}{a^2}(a^2 - x^2) dx = \frac{\pi b^2}{a^2}\left(a^2 x - \frac{x^3}{3}\right) + C.$$

Estimating the volume from the plane through the centre perpendicular to the transverse axis, we have $x = 0$, $C = 0$, and

$$v' = \frac{\pi b^2}{a^2}\left(a^2 x - \frac{x^3}{3}\right).$$

Making $x = a$, we obtain, for one-half the volume,

$$v'' = \frac{\pi b^2}{a^2}\left(a^3 - \frac{a^3}{3}\right) = \frac{2}{3}\pi b^2 a;$$

and for the whole,

$$\frac{4}{3}\pi b^2 a = \frac{2}{3}\pi b^2 \times 2a;$$

or, *equal to two-thirds of the circumscribing cylinder.*

If the same ellipse revolves about its conjugate axis, we have

$$v = \int \pi x^2 dy = \int \pi \frac{a^2}{b^2}(b^2 - y^2)dy,$$

which, between the limits $y = -b$, and $y = b$, gives

$$v'' = \frac{4}{3}\pi a^2 b = \frac{2}{3}\pi a^2 \times 2b.$$

The latter volume is called the *oblate spheroid*, and the former the *prolate spheroid;* and we have the proportion

$$\text{the prolate : the oblate} :: \frac{4}{3}\pi b^2 a : \frac{4}{3}\pi a^2 b :: b : a.$$

If in either expression $a = b$, we have

$$\frac{4}{3}\pi a^3 = \textit{volume of a sphere.}$$

Let the origin be now taken at A, when

$$y^2 = \frac{b^2}{a^2}(2ax - x^2),$$

and the volume be determined.

Give also the cubature of a sphere directly, by using the equation

$$y^2 + x^2 = \mathrm{R}^2.$$

256. Give also the cubatures of the following volumes of revolution:

. The right cone, $v'' =$ base $\times$ $\frac{1}{3}$ of altitude.

2. The paraboloid, $v'' = \frac{1}{2}$ circumscribing cylinder.

3. The volume generated by a given portion of the common parabola revolving about the tangent at its vertex,

$$v'' = \tfrac{1}{5} \text{ cylinder with same base and altitude.}$$

4. The volume, the bounding surface of which is generated by the curve whose equation is $y^2 = \frac{a}{x}$.

5. The volume, the bounding surface of which is generated by one branch of the cycloid revolving about its base.

257. II. *Of volumes bounded by any surface.* We have found in article (151), for the partial differential of a volume limited by a surface and the co-ordinate planes, the expression

$$d^2v = z\,dx\,dy \ldots\ldots\ldots\ldots (1).$$

To obtain an expression for the volume, we have simply to deduce from the equation of the bounding surface the value of z in terms of x and y, substitute it in (1), and then take the integral between proper limits. The result will be an expression for a definite portion of the volume. To indicate the process, we place equation (1) under the form

$$d\frac{dv}{dx} = z\,dy.$$

Integrating with respect to y,

$$\frac{dv}{dx} = \int z\,dy + \mathrm{X}.$$

From this,

$$dv = dx\int z\,dy + \mathrm{X}dx.$$

Integrating with reference to x,

$$v = \int dx\int z\,dy + \int \mathrm{X}dx;$$

or, Art. (196),

$$v = \int^2 z\,dx\,dy + \int \mathrm{X}dx + \mathrm{Y}.$$

The integral $\int z\,dy + \mathrm{X}$ is evidently the area of one of the parallel sections $d\mathrm{M}d'$, Art. (239). To obtain the whole volume represented in the figure, we must first take the integral between the limits $y = 0$, and $y = bd'$, this value of y being that deduced in terms of x from the equation of the curve $\mathrm{Y}d'\mathrm{X}$, and then the second integral between the limits $x = 0$, and $x = \mathrm{AX}$.

To illustrate, let us determine the volume of the pyramid ABD-C; the equation of the plane BDC, being

$$x + 2y + 3z - 2 = 0;$$

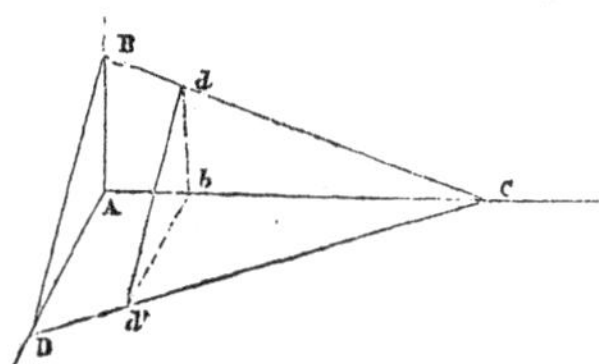

whence

$$z = \frac{2 - 2y - x}{3}.$$

The equation of DC is

$$x + 2y = 2, \qquad \text{or} \qquad y = 1 - \frac{x}{2},$$

$$AD = 1, \qquad AC = 2, \qquad AB = \frac{2}{3},$$

$$v = \int^2 z dx dy = \int dx \int dy \frac{(2 - 2y - x)}{3}$$

Integrating with respect to y,

$$v = \int dx \frac{2y - y^2 - xy}{3} + X;$$

or, taking the integral between the limits

$$y = 0, \qquad \text{and} \qquad y = bd' = 1 - \frac{x}{2},$$

$$v = \int dx \frac{1 - x + \frac{x^2}{4}}{3} = \frac{x - \frac{x^2}{2} + \frac{x^3}{12}}{3} + C.$$

Taking this between the limits

$$x = 0, \qquad \text{and} \qquad x = AC = 2,$$

we obtain for the volume,

$$v'' = \frac{4}{18} = \frac{1}{2} \times \frac{2}{3} \times 1 \times \frac{2}{3} = \frac{1}{2} AB \times AD \times \frac{1}{3} AC$$

$$= BAD \times \frac{1}{3} AC.$$

258. As the first integral with respect to y will often be complicated, it will be better, if possible, to obtain directly an expression for the area of the parallel section as dMd', in terms of x, multiply this by dx, and then integrate between the proper limits $x = 0$, and $x = AX$. Thus, for the elliptical paraboloid (see Analyt. Geometry) whose equation is

$$ny^2 + pz^2 = m''x,$$

by making first $y = 0$, and then $z = 0$, we have, for any section at a distance x from the plane YZ,

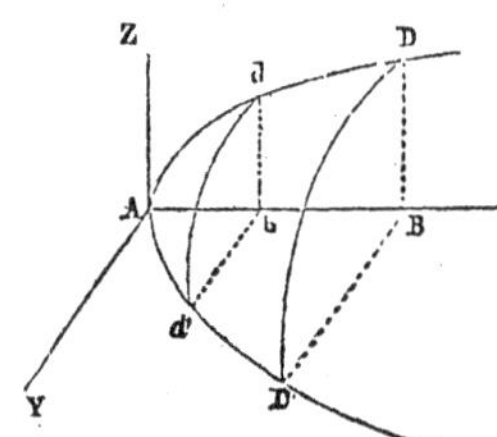

$$z = \sqrt{\frac{m''x}{p}} = bd,$$

$$y = \sqrt{\frac{m''x}{n}} = bd',$$

and for the area of the entire section,

$$\pi bd \times bd' = \pi\frac{m''x}{\sqrt{np}}.$$

Multiplying this by dx, we have

$$dv = \pi\frac{m''xdx}{\sqrt{np}}, \qquad \text{and} \qquad v = \frac{\pi m''x^2}{2\sqrt{np}} + C.$$

Taking the integral between the limits $x = 0$, and $x = AB = h$, we have

$$v'' = \frac{\pi m''h^2}{2\sqrt{np}} = \frac{\pi m''h}{\sqrt{np}} \times \frac{h}{2}.$$

The first factor of this is the area of the entire ellipse DBD′; hence, the volume of the paraboloid is equal to *half that of the circumscribing cylinder.*

In like manner, it may be shown that the volume of an ellipsoid whose equation is

$$b^2c^2x^2 + a^2c^2y^2 + a^2b^2z^2 = a^2b^2c^2,$$

is equal to *two-thirds that of the circumscribing cylinder.*

PART III.

CALCULUS OF VARIATIONS.

First Principles.

259. A function may be regarded as given, when the form of the algebraic expression, which determines the relation between it and the variable or variables, is given, and the constants which enter this expression are known.

In this case, the only change which the function can be made to undergo, is that which arises from a change in the variables. When these variables receive infinitely small increments, the corresponding infinitely small increment or change of the function is taken for *the differential of the function*, Art. (88). All our previous applications of the Calculus have been made to functions of the kind above referred to, and the term differential can, with propriety, be applied to no other change.

It will at once be seen, that if a function be not given as above described, but *merely subjected to certain conditions*, it may be made to undergo a change by altering the relation which exists between it and the variables; and this may be done by changing either the form of the expression for the function, or the constants which enter it, in any way consistent with the given conditions. Now, if such a change be made as to give *another function consecutive with the first*, the infinitely small change which the first undergoes is called *its variation*, and the corresponding changes of the variables are their *variations*.

The difference between the terms "differential" and "variation," will be made more plain by geometrical illustration.

Let BC be any curve, a function of x, Art. (90), of which M and M′ are any two consecutive points, the co-ordinates of M being x and y. Now, if the constants which determine the curve be changed in any way so as to give a different curve B′C′, *infinitely near* to BC, and so that the points M and M′ shall take the positions m and m', Pp will be the variation of x, and mS the variation of y, while PP′ is the differential of x, and M′Q the differential of y, Art. (88).

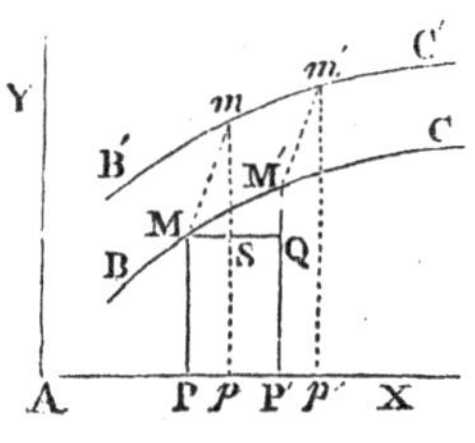

The conditions under which the variation is made, may be such that one of the variables will have no variation; and when this is the case, the operations to be performed will be much simplified.

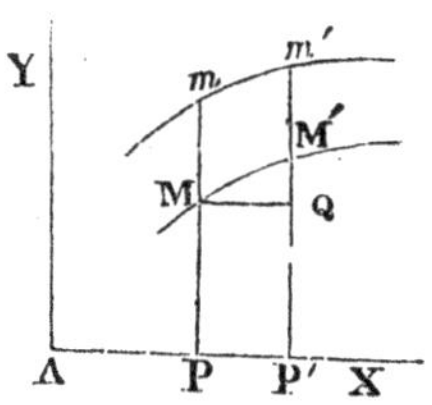

Thus, if it be required that the points M and M′ shall be found in lines parallel to the axis of Y at m and m', Mm will be the variation of y, while x has no variation; the differentials of x and y being PP′ and M′Q, as before.

As the differential is denoted by the symbol d, the Greek character δ is used to denote the variation; and from the illustrations just given, it appears that while the former symbol denotes the changes which take place in passing from *one point to another of the same curve*, the latter is used for a very different purpose, to denote the changes in passing from points of one curve to the corresponding points of another *infinitely near to it*.

260. From the nature of the term as above explained, we see that to obtain the variation of any function of x, y, z, &c., we have only to put for x, y, z, &c., $x + \delta x$, $y + \delta y$, &c.; δx, δy, &c., being, not arbitrary, but the infinitely small changes which take place in x, y, z, &c., in consequence of that change in the function which gives its variation; and then take, as in the Differential Calculus, Art. (52), those terms of the development which are of the first degree with reference to the variations of the variables. Or, since the development may be made precisely as in Art. (51), by substituting δx, δy, &c., for h, k, &c., it is plain that we shall have

$$\delta u = \frac{du}{dx}\delta x + \frac{du}{dy}\delta y + \frac{du}{dz}\delta z + \text{\&c.}$$

It is also plain that the principles contained in articles (15) and (17), as also the particular rules demonstrated in articles (20) (26), are equally applicable to variations.

261. In the function

$$u = f(x) \ldots\ldots\ldots\ldots (1),$$

let us substitute $x + \delta x$ for x, and denote the new function by $f'(x)$; then, by the definition, Art. (259),

$$\delta u = f'(x) - f(x) \ldots\ldots (2);$$

and since, from the relation expressed in equation (1), x is a function of u, the second member of equation (2) will be a function of u, and we may write

$$\delta u = \varphi(u) \ldots\ldots\ldots\ldots (3).$$

If, in this equation, we put for u, $u + du = u'$, we shall have

$$\delta u' = \varphi(u'),$$

and, subtracting equation (3),

$$\delta u' - \delta u = \varphi(u') - \varphi(u) = d\varphi(u) = d\delta u.$$

Taking the variation of the expression

$$u' - u = du,$$

we have

$$\delta u' - \delta u = \delta du;$$

hence

$$\delta du = d\delta u \ldots\ldots\ldots\ldots (4).$$

That is, *the variation of the differential of a function of a single variable is equal to the differential of its variation.* Or, when both of the symbols d and δ are prefixed to a function, the order in which they are written, or in which the operations indicated are performed, can be changed at pleasure without affecting the result.

The principle above enunciated is true for any order of the differential; for if, in equation (4), we put du for u, we have

$$\delta d(du) = d\delta du, \qquad \text{or} \qquad \delta d^2 u = dd\delta u = d^2\delta u.$$

If, in the last equation, we put du for u, we have

$$\delta d^2(du) = d^2\delta du, \qquad \text{or} \qquad \delta d^3 u = d^3\delta u,$$

and so on; hence we may conclude that

$$\delta d^n u = d^n \delta u.$$

262. Let v be any differential of a function of x, and place

$$\int v = v', \qquad \text{then} \qquad dv' = v,$$

$$\delta dv' = \delta v, \qquad \text{or} \qquad d\delta v' = \delta v,$$

and by integration,

$$\delta v' = \int \delta v, \qquad \text{or} \qquad \delta \int v = \int \delta v.$$

The principles demonstrated in this and the preceding article, are evidently true for functions of any number of variables; since the variation of the differential of such a function is but the sum of the partial variations, and the converse.

263. In order to consider the subject of variations in its most general sense, when applied to differential expressions, we must regard the differentials of all the variables as variable, as well as the variables themselves. In this sense, if u be a function containing x, y, and their successive differentials, we shall have, Art. (260),

$$\left.\begin{aligned} \delta u = {} & \mathrm{M}\delta x + \mathrm{M}'\delta dx + \mathrm{M}''\delta d^2x + \&c. \\ & + \mathrm{N}\delta y + \mathrm{N}'\delta dy + \mathrm{N}''\delta d^2y + \&c. \end{aligned}\right\}\ldots(1),$$

in which M, M′, M″, &c., are the partial differential coefficients of u taken with respect to x, dx, d^2x, &c.; and N, N′, N″, &c., the corresponding ones taken with respect to y, dy, d^2y, &c. This expression may be extended to any number of variables, by adding for each, an expression of the form

$$\mathrm{M}\delta x + \mathrm{M}'\delta dx + \mathrm{M}''\delta d^2x + \&c.;$$

and may then be made to give every particular case which can arise, by making the particular suppositions upon dx, d^2x, dy, d^2y, &c., which the case requires.

264. If the differential expression contains only the variables x, y, $\frac{dy}{dx} = p$, $\frac{d^2y}{dx^2} = q$, &c., we may denote it by v, and shall have, as in Art. (260),

$$\delta v = M\delta x + N\delta y + N'\delta p + N''\delta q + \&c. \ldots (2).$$

And if this expression be taken in its most general sense, dx must be regarded as variable; in which case, we put for δp, δq, &c., their values obtained as in Art. (26), viz.:

$$\delta p = \delta \frac{dy}{dx} = \frac{dx\delta dy - dy\delta dx}{dx^2} = \frac{d\delta y - pd\delta x}{dx},$$

$$\delta q = \delta \frac{dp}{dx} = \frac{dx\delta dp - dp\delta dx}{dx^2} = \frac{d\delta p - qd\delta x}{dx}.$$

If dx be regarded as constant, equation (2) is under its most simple form.

265. If we indicate the integration of both members of equation (1), Art. (263), we have

$$\left.\begin{aligned} \int \delta u &= \int (M\delta x + M'\delta dx + M''\delta d^2x + \&c.) \\ &+ \int (N\delta y + N'\delta dy + N''\delta d^2y + \&c.) \end{aligned}\right\} \ldots (1).$$

By the application of the rule for integrating by parts, we find

$$\int M'\delta dx = \int M'd\delta x = M'\delta x - \int dM'\delta x;$$

$$\begin{aligned} \int M''\delta d^2x &= \int M''d^2\delta x = M''d\delta x - \int dM''d\delta x \\ &= M''d\delta x - dM''\delta x + \int d^2M''\delta x; \end{aligned}$$

$$\begin{aligned} \int M'''\delta d^3x &= \int M'''d^3\delta x = M'''d^2\delta x - \int dM'''d^2\delta x \\ &= M'''d^2\delta x - dM'''d\delta x + \int d^2M'''d\delta x \\ &= M'''d^2\delta x - dM'''d\delta x + d^2M'''\delta x - \int d^3M'''\delta x. \end{aligned}$$

Also,

$$\int N'\delta dy = N'\delta y - \int dN'\delta y;$$

$$\int N''\delta d^2y = N''d\delta y - dN''\delta y + \int d^2N''\delta y;$$

$$\int N'''\delta d^3y = N'''d^2\delta y - dN'''d\delta y + d^2N'''\delta y - \int d^3N'''\delta y.$$

Observing that the second member of equation (1) is equal to the sum of the integrals of the terms taken separately, and substituting the above values, we obtain

$$\begin{aligned}\int\delta u = (M' - dM'' + d^2M''' - \&c.)\delta x + (M'' - dM''' + \&c.)d\delta x \\ + (M''' - \&c.....)d^2\delta x + \&c. \\ + (N' - dN'' + d^2N''' - \&c.)\delta y + (N'' - dN''' + \&c.)d\delta y \\ + (N''' - \&c.....)d^2\delta y + \&c. \\ \left.\begin{aligned} + \int(M - dM' + d^2M'' - d^3M''' + \&c.)\delta x \\ + \int(N - dN' + d^2N'' - d^3N''' + \&c.)\delta y \end{aligned}\right\} \quad \ldots\ldots\ldots (2).\end{aligned}$$

By examining the above expression, it will be seen that there is no term under the sign $\int$ which contains the symbols d and δ applied the one to the other; and also that the parts containing δx are exactly similar to those containing δy. The formula may therefore be extended to any number of variables, by adding, for each new variable, similar parts containing its variation.

266. It should be remarked, that if the multipliers of δx and δy following the sign $\int$, in equation (2) of the preceding article, are both equal to zero, $\int\delta u$ will be complete, or δu will be the differential of some function. But in the expression

$$\int\delta u = \delta\int u,$$

it is evident that if $\int u$ contain any terms which cannot be freed from the sign $\int$, $\delta\int u$ must contain the variations of these terms still under the sign, and $\int \delta u$ cannot be complete. Hence, if δu is a differential, u itself must be so. And conversely; for if $\int u$ is entirely freed from the sign $\int$, then $\delta\int u$ cannot contain this sign, and its equal $\int\delta u$ must be complete, or δu be a differential. Hence, if the conditions

$$M - dM' + d^2M'' - \&c. = 0,$$

$$N - dN' + d^2N'' - \&c. = 0,$$

are satisfied, u will be the differential of some function, which may be obtained by integration. If the above conditions are not satisfied, u cannot be an exact differential, and $\int u$ cannot be obtained.

267. If we take the variation of the expression $\int vdx$, in which v, as in Art. (264), is a function of x, y, p, q, &c., we have, Arts. (21) and (155),

$$\delta\int vdx = \int\delta(vdx) = \int v\delta dx + \int dx\delta v.$$

But, Art. (169),

$$\int v\delta dx = \int vd\delta x = v\delta x - \int dv\delta x;$$

hence

$$\delta\int vdx = v\delta x + \int(dx\delta v - dv\delta x)\dots\dots.(1).$$

Substituting in that part of the second member which follows the sign $\int$, the values of dv and δv, Arts. (52) and (264),

$$dv = Mdx + Ndy + N'dp + N''dq + \&c.,$$

$$\delta v = M\delta x + N\delta y + N'\delta p + N''\delta q + \&c.,$$

we have

$$dx\delta v - dv\delta x = \text{N}(dx\delta y - dy\delta x) + \text{N}'(dx\delta p - dp\delta x)$$

$$+ \text{N}''(dx\delta q - dq\delta x) + \&\text{c.} \ldots\ldots\ldots\ldots (2).$$

Since $dy = pdx$, we have

$$dx\delta y - dy\delta x = dx(\delta y - p\delta x) = \omega dx,$$

by making $\delta y - p\delta x = \omega$.

Also, if for δp, we put its value, Art. (264), we have

$$dx\delta p - dp\delta x = d\delta y - pd\delta x - dp\delta x = d(\delta y - p\delta x) = d\omega.$$

If, in this last expression, we put p for y, and q for p, and recollect that $q = \dfrac{dp}{dx}$, we have

$$dx\delta q - dq\delta x = d(\delta p - q\delta x) = d\left(\frac{dx\delta p - dp\delta x}{dx}\right) = d\left(\frac{d\omega}{dx}\right).$$

Substituting these values in equation (2), and prefixing the sign $\int$, we have

$$\int(dx\delta v - dv\delta x) = \int \text{N}\omega dx + \int \text{N}'d\omega + \int \text{N}''d\left(\frac{d\omega}{dx}\right) + \&\text{c.} \ldots (3).$$

Again, by Art. (169),

$$\int \text{N}'d\omega = \text{N}'\omega - \int \frac{d\text{N}'}{dx}\omega dx,$$

$$\int \text{N}''d\frac{d\omega}{dx} = \text{N}''\frac{d\omega}{dx} - \int \frac{d\text{N}''}{dx}d\omega$$

$$= \text{N}''\frac{d\omega}{dx} - \frac{d\text{N}''}{dx}\omega + \int \frac{1}{dx}d\left(\frac{d\text{N}''}{dx}\right)\omega dx.$$

Now, substituting these expressions in (3), and the result in (1), we obtain

$$\delta \int v dx = v\delta x + (\mathrm{N}' - \frac{d\mathrm{N}''}{dx} + \&c.)\,\omega + (\mathrm{N}'' - \&c.)\frac{d\omega}{dx} + \&c.$$

$$+ \int (\mathrm{N} - \frac{d\mathrm{N}'}{dx} + \frac{1}{dx} d\frac{d\mathrm{N}''}{dx} - \&c.)\,\omega dx \ldots . (4).$$

If we now put for ω, its value $\delta y - p dx$, the part affected with the sign $\int$ will become

$$\int (\mathrm{N} - \frac{d\mathrm{N}'}{dx} + \&c.)\, dx\delta y \; - \int (\mathrm{N} - \frac{d\mathrm{N}'}{dx} + \&c.)\, p dx \delta x.$$

From which we see that, in this case, the coefficients of δy and δx have such a relation that if one becomes equal to zero the other will.

Maxima and Minima of Indeterminate Integrals.

268. The principal, and far the most important application of variations, is to the determination of the *maxima* and *minima* of indeterminate integrals, that is, of integral expressions of the form

$$\int \sqrt{dx^2 + dy^2}, \qquad \int \pi y^2 dx, \qquad \&c.,$$

containing x, y, &c., and their differentials, in which the relation between the variables is entirely unknown. Thus, if it be required to determine the relation between x and y, in order that $\int \pi y^2 dx$ taken under certain conditions, shall be a maximum or minimum, the problem is one not capable of solution by the ordinary method of article (69), since the principles there developed require the form of the function to which they are to be applied, and the constants which enter it, to be given, and the search is for particular values of the variables, which will make one or more values of the function a maximum or minimum; whereas the object now proposed, is to ascertain what this form and these constants must be, in order that the function, when subjected to the given conditions, shall be a maximum or minimum, the variables being entirely indeterminate. Questions of this kind are readily solved by the aid of variations.

269. Let u be a function of the nature discussed in Art. (263), and suppose x, dx, y, dy, &c., to be increased by their variations; and let the difference between the corresponding function u' and u be developed, which is done at once by putting δx, δy, δdx, &c., for h, k, l, &c., in the development of Art. (51); we shall thus obtain

$$u' - u = \mathrm{M}\delta x + \mathrm{N}\delta y + \mathrm{M}'\delta dx + \mathrm{N}'\delta dy + \&c.,$$

plus a term of the second degree with respect to δx, δy, &c.; plus other terms.

By the same course of reasoning as that contained in Art. (77), we see that u can be neither greater nor less than u', for all values of δx, δy, &c., unless the term, of the first degree with reference to these variations, is equal to zero. But this term, Art. (263), is the variation of u: *Hence, in order that u be a maximum or minimum, δu must be equal to zero.*

If the conditions which make the variation of u equal to zero, make the term of the second degree, in the above development, positive, for all values of δx, δy, &c., u will be a minimum; if negative, u will be a maximum. The discussion of the various circumstances in which this term will not change its sign, is of too complicated a nature, and likely to lead too far, for an elementary treatise. Neither is it necessary in general, as we shall be able, from the nature of nearly every case, to determine, without a reference to this second term, whether we have a maximum or minimum.

270. In the application of the foregoing principles to the indeterminate integrals referred to in Art. (268), it may at first be remarked, that if the integral be *indefinite*, Art. (160), from its nature it can have no maximum nor minimum. The application can then only be made to *definite integrals*, or those which are taken between some well-defined limits.

If, then, it be required that $\int u$ be a maximum or minimum, we may write the variation of $\int u$, Art. (265), thus:

$$\delta \int u = \int \delta u = m\delta x + n\delta y + m'\delta dx + n'\delta dy + \&c.,$$

$$+ \int (k\delta x + k'\delta y) \ldots\ldots\ldots\ldots (1);$$

and this, when taken between the prescribed limits, must be equal to zero.

We have seen, Art. (266), that this expression cannot be integrated unless the quantity which follows the sign $\int$, in the second member, is equal to zero; that is, there can be no integral to be taken between limits, and of course no maximum nor minimum. We must then have, for the first condition,

$$k\delta x + k'\delta y = 0 \ldots\ldots\ldots\ldots (2).$$

If, in the particular case under discussion, the variations of x and y are entirely independent of each other, we must also have

$$k = 0, \qquad \text{and} \qquad k' = 0;$$

or, Art. (265),

$$\left.\begin{array}{l} \mathrm{M} - d\mathrm{M}' + d^2\mathrm{M}'' - \&c. = 0 \\ \mathrm{N} - d\mathrm{N}' + d^2\mathrm{N}'' - \&c. = 0 \end{array}\right\} \ldots (3).$$

Again, if we denote by l and l' the results obtained by substituting the limits in succession in the remaining part of equation (1), we must have, for a second condition,

$$l' - l = 0 \ldots\ldots\ldots\ldots (4).$$

Should there be more than two variables in the function u, the quantity following the sign $\int$, in equation (1), will consist of as many terms as there are variables, each of which, if the variations are independent of each other, must be placed equal to zero, and will thus give an equation expressing a relation between these variables and their differentials.

If, however, the conditions under which the variations are made are such as to render these variations in any way dependent, we shall be able, by means of the equations which express these conditions, to eliminate from equation (1) one or more of these variations; then, by placing the coefficients of those which remain under the sign $\int$, equal to zero, we shall have a system of equations from which we may determine the nature and extent of the required function. The system of equations (3) will, in every case, express the relation which must exist between the variables and their differentials, in order that the function shall be a maximum or minimum; but they must be subjected to the conditions deduced from the equation

$$l' - l = 0,$$

which can, of course, contain no variables except those which belong exclusively to the limits.

Where u is under the form vdx, it has been seen, Art. (267), that the two equations (3) will both be satisfied, if one is. They will therefore give but one independent equation, viz.:

$$N - \frac{dN'}{dx} + \frac{1}{dx} d \frac{dN''}{dx} - \&c. = 0 \ldots . (4);$$

and the condition

$$l' - l = 0$$

must be deduced by substituting the limits in that part of equation (4), Art. (267), which is independent of the sign $\int$.

The solution and discussion of the following problems will serve to illustrate and more fully develop the preceding principles.

271. *Problem* 1.—Required the nature of the shortest line joining two given points in a plane.

Let x', y', and x'', y'', be the co-ordinates of the points. The general expression for the length of the line, Art. (234), is

$$z = \int \sqrt{dx^2 + dy^2}.$$

Taking the variation of this, we have

$$\delta \int u = \int \left(\frac{dx \delta dx}{dz} + \frac{dy \delta dy}{dz} \right);$$

which, upon comparison with equation (1), Art. (263), gives

$$\text{M} = 0, \qquad \text{N} = 0, \qquad \text{M}' = \frac{dx}{dz}, \qquad \text{N}' = \frac{dy}{dz},$$

and all the other terms equal to zero. In this case, since δx and δy are independent of each other, we use equations (3) of the preceding article, and have

$$d\frac{dx}{dz} = 0, \qquad \text{and} \qquad d\frac{dy}{dz} = 0;$$

whence, by integration,

$$\frac{dx}{dz} = c, \qquad \frac{dy}{dz} = c'.$$

Eliminating dz, and integrating again, we have

$$dy = \frac{c'}{c} dx = a dx, \qquad y = ax + b \ldots . (1),$$

which gives the required relation between y and x, and indicates that the line must be straight.

The first part of equation (2), Art. (265), becomes

$$M'\delta x + N'\delta y.$$

Since, in this case, the limits x', y', and x'', y'', are absolutely fixed, we must have $\delta x'$, $\delta y'$, &c., equal to zero, which, being substituted in the above expression, give

$$M'\delta x' + N'\delta y' = 0, \qquad M'\delta x'' + N'\delta y'' = 0;$$

whence results the fulfilment of the second condition,

$$l' - l = 0,$$

and it remains only to determine the constants a and b, in equation (1), on condition that the line shall pass through the two given points.

272. *Problem* 2.—Required the shortest line that can be drawn from one given curve to another, in the same plane.

Let $\qquad y = f(x), \qquad$ and $\qquad y = f'(x),$

be the equations of the curves; their differential equations being

$$dy = p'dx, \qquad dy = p''dx \ldots\ldots (1).$$

As in the preceding problem, we have

$$z = \int\sqrt{dx^2 + dy^2}, \qquad \delta\int u = \int\left(\frac{dx}{dz}\delta dx + \frac{dy}{dz}\delta dy\right);$$

from which is deduced, precisely as before, the equation of the required line,

$$y = ax + b \ldots\ldots\ldots\ldots (2).$$

But since the ends of this line must be in the given curves, the variations of x and y, at the limits, must be confined to these curves, that is, $\delta y'$, $\delta x'$, $\delta y''$, $\delta x''$ must be the same as dy and dx in equations (1); whence

$$\delta y' = p'\delta x', \qquad \delta y'' = p''\delta x''.$$

Substituting these, in succession, in the first part of equation (2), Art. (265), and subtracting the results, we must have

$$l' - l = \left(\frac{dx'}{dz'} + \frac{dy'}{dz'}p'\right)\delta x' - \left(\frac{dx''}{dz''} + \frac{dy''}{dz''}p''\right)\delta x'' = 0;$$

and since this contains two independent variations, it can only be satisfied by making the coefficients separately equal to zero; hence

$$dx' + dy'p' = 0, \qquad dx'' + dy''p'' = 0;$$

whence

$$\frac{dy'}{dx'} = -\frac{1}{p'}, \qquad \frac{dy''}{dx''} = -\frac{1}{p''}.$$

But these are the equations of condition that the required line shall be normal to both curves at the points (x', y'), (x'', y''), respectively, Art. (84).

In order to determine the constants a and b, in equation (2), we must first find the values of x', y', x'', y'', on condition that the normal to the first curve at the point (x', y') shall also be normal to the second at the point (x'', y''), and then cause the line to pass through these points.

This problem and the preceding may also be solved by placing

$$z = \int\sqrt{dx^2 + dy^2} = \int\left(1 + \frac{dy^2}{dx^2}\right)^{\frac{1}{2}} dx = \int v dx,$$

in which v is a function of $\frac{dy}{dx} = p$. In this case, we should use equations (4), Arts. (267) and (270).

273. *Problem* 3.—Required the shortest line, on the surface of a sphere, joining two given points of the surface.

Let the equation of the sphere be

$$x^2 + y^2 + z^2 = R^2 \ldots\ldots\ldots\ldots (1).$$

The general expression for the length of a line joining the two points will be, Art. (91),

$$s = \int \sqrt{dx^2 + dy^2 + dz^2},$$

the variation of which is

$$\delta \int u = \int \left(\frac{dx}{ds} \delta dx + \frac{dy}{ds} \delta dy + \frac{dz}{ds} \delta dz \right);$$

whence, by adding an expression containing δz to the second member of equation (2), Art. (263), and comparing, we find

$$M = 0, \qquad N = 0, \qquad P = 0,$$

$$M' = \frac{dx}{ds}, \qquad N' = \frac{dy}{ds}, \qquad P' = \frac{dz}{ds},$$

and all the other terms equal to 0. The first condition required in Art. (270), is then

$$d\left(\frac{dx}{ds}\right) \delta x + d\left(\frac{dy}{ds}\right) \delta y + d\left(\frac{dz}{ds}\right) \delta z = 0 \ldots . (2).$$

In this case the variations are not independent, but must be confined to the surface of the sphere; that is, taking the variation of equation (1), we must have

$$2x\delta x + 2y\delta y + 2z\delta z = 0.$$

Combining this with equation (2), and eliminating δz, we obtain

$$\left(d\frac{dx}{ds} - \frac{x}{z}d\frac{dz}{ds}\right)\delta x + \left(d\frac{dy}{ds} - \frac{y}{z}d\frac{dz}{ds}\right)\delta y = 0,$$

which, containing two independent variations, gives

$$zd\frac{dx}{ds} - xd\frac{dz}{ds} = 0, \qquad zd\frac{dy}{ds} - yd\frac{dz}{ds} = 0.$$

Now, if we regard ds as constant, these equations become

$$z\frac{d^2x}{ds} - x\frac{d^2z}{ds} = 0, \qquad z\frac{d^2y}{ds} - y\frac{d^2z}{ds} = 0,$$

from which we deduce

$$x\frac{d^2y}{ds} - y\frac{d^2x}{ds} = 0.$$

Integrating the last three equations, we have

$$z\frac{dx}{ds} - x\frac{dz}{ds} = a, \qquad z\frac{dy}{ds} - y\frac{dz}{ds} = b, \qquad x\frac{dy}{ds} - y\frac{dx}{ds} = c.$$

Multiplying the first by y, the second by $-x$, the third by z, and adding, we obtain

$$ay - bx + cz = 0 \ldots\ldots.(3),$$

which is the equation of a plane passing through the centre of the sphere. The required curve must lie in this plane, and therefore is the arc of a great circle.

The limits in this case, as in problem 1, being absolutely fixed, we have at once, as in that problem, the fulfilment of the second condition,

$$l' - l = 0.$$

Equation (3) may be put under the form

$$\frac{a}{c}y - \frac{b}{c}x + z = 0, \qquad \text{or} \qquad z = c'x + d'y,$$

and the constants c' and d' determined, by causing the plane to pass through the given points.

274. In many cases where there are conditions confining the variations, whether at the limits or not, the method of reducing the number of independent variations, explained in Art. (270), and pursued in Arts. (272, 273), will be found of very difficult application. In all these cases, the following less direct, but very elegant method may be used. Let

$$r = 0, \qquad s = 0, \qquad \text{\&c.},$$

be the equations between x, y, &c., expressing the conditions to which the variations are subject; then, at the same time that we have

$$\delta \int u = 0,$$

we must also have

$$\delta r = 0, \qquad \delta s = 0, \qquad \text{\&c.};$$

or, denoting by c, c', &c., arbitrary constants, we must have the equation

$$\delta \int u + c\delta r + c'\delta s + \text{\&c.} = 0 \ldots\ldots(1),$$

for all values of the variations of x, y, &c. Placing the coefficients of these variations separately equal to zero, we obtain equations from which we can eliminate the constants c, c', &c., and thus deduce an equation or equations which will express the proper relation between x, y, &c. As an illustration, let us take

Problem 4.—Required the nature of the line, of a given length, joining two points, which, with the ordinates of the points and axis of X, will inclose the greatest area. In this case we have, Art. (239),

$$\delta\int u = \delta\int y dx;$$

and since the length of the arc between the limits is to be constant, the variations must be subject to the condition

$$\int dz = \int\sqrt{dx^2 + dy^2} = a;$$

hence

$$\delta\int\sqrt{dx^2 + dy^2} = 0.$$

Equation (1) will then become

$$\delta\int y dx + c\delta\int\sqrt{dx^2 + dy^2} = 0;$$

or, putting for the variations their values, we have

$$\int\left(y\delta dx + dx\delta y + \frac{cdx\delta dx + cdy\delta dy}{dz}\right) = 0.$$

Comparing this with equation (1), Art. (263), we see that

$$\text{M} = 0, \quad \text{M}' = y + c\frac{dx}{dz}, \quad \text{N} = dx, \quad \text{N}' = c\frac{dy}{dz};$$

and these, being substituted in equations (3), of Art. (270), give

$$-\,d\left(y + c\frac{dx}{dz}\right) = 0, \quad dx - cd\left(\frac{dy}{dz}\right) = 0;$$

and by integrating,

$$y + c\frac{dx}{dz} = \beta, \qquad x - c\frac{dy}{dz} = \alpha.$$

Eliminating c from these two equations, we obtain

$$\frac{dy}{dx} = -\frac{x - \alpha}{y - \beta},$$

which is evidently the differential equation of a circle whose equation is, Art. (98),

$$(y - \beta)^2 + (x - \alpha)^2 = R^2,$$

β, α, and R being arbitrary constants, which must be determined on condition that the circle pass through the two given points, and that the included arc be of the given length.

www.ingramcontent.com/pod-product-compliance
Lightning Source LLC
LaVergne TN
LVHW022039110826
845151LV00007B/193

* 9 7 8 1 4 2 5 5 4 1 9 7 2 *